Just Food

Just Food

Philosophy, Justice and Food

J. M. Dieterle

ROWMAN &
LITTLEFIELD
INTERNATIONAL

London • New York

Published by Rowman & Littlefield International, Ltd.
Unit A, Whitacre Mews, 26–34 Stannary Street, London SE11 4AB
www.rowmaninternational.com

Rowman & Littlefield International, Ltd. is an affiliate of Rowman & Littlefield
4501 Forbes Boulevard, Suite 200, Lanham, Maryland 20706, USA
With additional offices in Boulder, New York, Toronto (Canada), and Plymouth (UK)
www.rowman.com

British Library Cataloguing in Publication Data
A catalogue record for this book is available from the British Library

ISBN: HB 978-1-7834-8386-0
PB 978-1-7834-8387-7

Library of Congress Cataloging-in-Publication Data
Just food: philosophy, justice, and food / edited by J. M. Dieterle.
pages cm
Includes bibliographical references and index.
ISBN 978-1-78348-386-0 (cloth: alk. paper)—ISBN 978-1-78348-387-7 (pbk.: alk. paper)—ISBN 978-1-78348-388-4 (electronic) 1. Food security. 2. Food supply—Moral and ethical aspects. I. Dieterle, Jill Marie, 1964-editor.
HD9000.5.J837 2015
363.8—dc23

2015028260

♾™ The paper used in this publication meets the minimum requirements of American National Standard for Information Sciences—Permanence of Paper for Printed Library Materials, ANSI/NISO Z39.48-1992.

Printed in the United States of America

Contents

Acknowledgements

Many people deserve thanks for their help with this project. First and foremost, I thank those who contributed their work to the volume. Without them, *Just Food: Philosophy, Justice and Food* would not be. I thank especially Lori Watson for her advice and editing suggestions, and Margaret Crouch, Nancy E. Snow and Paul B. Thompson for their guidance and encouragement.

I thank Eastern Michigan University for awarding me a Faculty Research Fellowship in the fall semester of 2014 to work on the volume. I also thank Sarah Campbell and Sinéad Murphy at Rowman and Littlefield International for all of their help with the project.

Sherri Switala deserves special thanks for motivating me to think about food and ethics some 30 years ago. Last, but certainly not least, I thank Eric Buckhave. His love, encouragement, and unwavering support kept me steady through each stage of this project, despite all of the challenges I encountered along the way. I could not have done it without him.

Introduction

J. M. Dieterle

The phrase "food justice" is often invoked to highlight a range of ethical issues concerning food, including but not limited to food production and the rights of agricultural laborers; inequities in food distribution within nations and between them; increased obesity rates among the poor and lack of access to healthy nutritious foods as the primary cause of such increases; the corporatization of the food system; the unsustainable nature of our current methods of food production; and the lack of democratic control over how food is grown, harvested and distributed. As, such "food justice" has emerged as an important organizing concept for those interested in the complex forms of inequality and injustice that permeate and sustain food systems across the globe.

Philosophical work on justice has a long and rich tradition, going back to Plato's Republic. However, much philosophical work on justice has focused narrowly on issues of states' authority to govern and what states owe their citizens. Of particular concern to political philosophers is the concept of distributive justice, which deals with the fair or just distribution of society's burdens and benefits. Within discussions of justice broadly, and distributive justice in particular, philosophers have largely treated food as just another resource to be distributed or ignored it altogether. Of course, there are some notable exceptions. Peter Singer has been writing about global famine and our duties to those living in absolute poverty since his seminal 1972 article, "Famine, Affluence, and Morality."[1] World Hunger has received much attention from philosophers, including James Sterba, Onora O'Neill, Thomas Pogge and Hugh LaFollette.[2] But philosophical work on the *particular* issues that arise with respect to food, including food access, food insecurity, systems of food production, the rights of workers in the agricultural context, the relations between nations concerning global food production and the disproportionate ways in which food insecurity impacts women and people of color are only recently getting systematic and full treatment by philosophers. *Just Food: Philosophy, Justice and Food* is a collection of essays by philosophers on these pressing issues of food justice. Here philosophers bring the rich history of philosophical thinking about justice to bear on issues of food justice. Who has access to food, and why? Who is denied such access, and why? What are the consequences of food insecurity? What values should drive food production? What would it take for a food system to be just?

THE ISSUES

The United Nations Declaration of Human Rights recognizes the right to food as a human right.[3] We currently produce more than enough food to feed the world, yet, globally, at least 842 million people (1 in 8) are malnourished or undernourished.[4] Although 98 percent of food insecure individuals live in developing countries, food insecurity is prevalent even in food-rich nations. The USDA estimates that 14.5 percent of households in the United States were food insecure at some point during 2012.[5]

Food security exists "when all people, at all times, have physical, social and economic access to sufficient, safe and nutritious food which meets their dietary needs and food preferences for an active and healthy life."[6] This definition points to a number of ways that individuals could be or become food insecure: they may not have physical access to food (food is not available, or it is not available on a consistent basis); there may be social or cultural reasons why they do not have sufficient food (perhaps gender norms dictate that the men in the household eat first); they may lack the economic resources to access food; or the available food may be unsafe or unhealthy. In line with this, the World Health Organization lists three facets of food security: (1) food availability (that a sufficient quantity of healthy, nutritious food is available on a consistent basis); (2) food access (that one has sufficient resources to access healthy, nutritious food); and (3) food use (that one has the knowledge of basic nutrition and sanitation to properly feed oneself).[7] Globally, poverty is the leading cause of hunger. The vast majority of those who are food insecure are so because they lack the resources to access healthy, nutritious food. Further, poverty and hunger reinforce each other. Chronic hunger impacts one's ability to function and, thus, one's ability to learn and work. As such, one's ability to undertake the tasks necessary to feed oneself is further compromised.

As noted above, we currently produce enough food to feed the world. If calories were distributed equally across the globe, at current levels of production we could provide each person with 2700 kcal—a sufficient number to maintain a healthy life.[8] However, calories are not distributed equally. Some live in food insecurity while others have access to and the means to purchase bountiful quantities of food. Further, much food goes to waste in the developed world. Thirty-one percent of food available for consumption in the United States in 2010 ended up in landfills;[9] 7 million tons of food is thrown away every year in the United Kingdom (more than half of which is still edible);[10] and the European Union as a whole annually wastes more than 100 million tons of food.[11]

The future is likely to bring considerable challenges. Global climate change poses a substantial threat to food production, especially in the global south. Energy needs are likely to result in increased demand for biofuels, and thus there is the potential for a significant portion of the world's arable land to be devoted to the production of such fuels instead of food. Further, world population is projected to grow by over 2.5 billion—reaching 9.6 billion—by 2050.[12] As per capita income grows in developing countries, the consumption of meat and other animal products typically increases. The production of meat requires far more of the earth's resources than the production of grains and vegetables. The United Nations Food and Agricultural Organization (FAO) estimates that given these considerations, agricultural output will have to increase by 60 percent over 2005–2007 levels to meet the 2050 demand.[13] In light of these challenges, attaining the goal of global food security may prove to be a formidable task.

Some argue that the way to fight food insecurity and attain food justice is through trade liberalization. If borders are open and trade regulations are relaxed or eliminated, then each nation can produce what it is "good at" producing and trade with other countries to get what it needs. According to those who defend this line of reasoning, trade is more *efficient* than each nation attempting to feed itself. C. Ford Runge and Benjamin Senauer concisely spell out the argument:

> The challenge of food security is . . . a race between productivity and populations with rising incomes. Here is where trade can make a difference. It enables food—primarily grain—to move from areas of surplus to areas of deficit, allowing the deficit regions to feed themselves as long as they can pay. Expanded access to rich-country markets also increases the export earnings of developing countries by raising the cash needed to buy food and other goods. Conversely, anything that restricts this movement or reduces the ability to pay for food imports will damage this capacity.[14]

Trade restrictions and tariffs, they argue, hamper trade and thus should be eliminated.

The World Bank, the International Monetary Fund and the World Trade Organization all embrace and promote the trade liberalization strategy. However, critics of trade liberalization are quick to point out that policies that promote unrestricted trade tend to concentrate food production in the hands of large agribusinesses and favor wealthier, more powerful nations. Further:

> Excess production is off-loaded through "dumping," an international trade strategy that places food in targeted export markets at prices below the cost of production. This practice has had devastating effects on domestic agricultural systems, which cannot compete with the influx of subsidized commodities saturating local markets.[15]

Peasant farmers are harmed when food is imported at prices below what it costs to grow it. For example, since the North American Free Trade Agreement (NAFTA) eased trade restrictions on trade between countries in North America, U.S. corn imports have pushed corn prices down in Mexico. Mexican farmers must now compete with subsidized U.S. corn, which has seriously compromised their ability to support themselves and their families; 1.3 million agricultural jobs have been lost in Mexico since NAFTA.[16] Haiti offers another example. Native-grown rice in Haiti could not compete with imported rice after trade liberalization policies were introduced. Many subsistence farmers left their farms to find work in factories. Haiti now imports 80 percent of its rice; in the 1980s, it was self-sufficient in rice production.[17]

If the goal is merely food *security,* then the emphasis is on adequate nutrition for all peoples. As a result, efficiency and (technologically driven) productivity are the values that drive policy. But much is left out if efficiency and productivity are one's primary values. The Food Sovereignty movement rejects the goals of and values behind the policies promoted by the World Bank, IFC and WTO. Food Sovereignty stresses the right of peoples to *define their own food system.* At the Forum for Food Sovereignty in 2007, the Declaration of Nyéléni was adopted by over 500 representatives from 80 different countries. In part, the declaration reads:

> Food sovereignty is the right of peoples to healthy and culturally appropriate food produced through ecologically sound and sustainable methods, and their right to define their own food and agriculture systems. It puts the aspirations and needs of those who produce, distribute and consume food at the heart of food systems and policies rather than the demands of markets and corporations. It defends the interests and inclusion of the next generation. It offers a strategy to resist and dismantle the current corporate trade and food regime, and directions for food, farming, pastoral and fisheries systems determined by local producers and users. Food sovereignty prioritises local and national economies and markets and empowers peasant and family farmer-driven agriculture, artisanal fishing,

> pastoralist-led grazing, and food production, distribution and consumption based on environmental, social and economic sustainability. Food sovereignty promotes transparent trade that guarantees just incomes to all peoples as well as the rights of consumers to control their food and nutrition. It ensures that the rights to use and manage lands, territories, waters, seeds, livestock and biodiversity are in the hands of those of us who produce food. Food sovereignty implies new social relations free of oppression and inequality between men and women, peoples, racial groups, social and economic classes and generations.[18]

Food Sovereignty is a bottom-up movement that aims to put control of the food supply in the hands of local producers and users.

In developed countries, food access is often hampered by the existence of food deserts: low-income communities without venues for purchasing fresh, healthy and affordable food. Although food deserts may have "food" available, it isn't healthy, nutritious food. Typically, food deserts have an array of fastfood restaurants and "fringe" food locations: gas stations, liquor stores, party stores and convenience stores.[19] Fringe food locations typically stock highly processed, packaged or canned food, much of which is high in fat, sugar and salt.[20] These foods have limited nutritional value, and studies have shown that there are significant health outcomes related to food deserts.[21]

Food injustices tend to track other injustices. Traditionally oppressed groups, such as communities of color and those living in poverty, are disproportionately affected by food insecurity. Globally, gender inequities impact women's access to food.[22] Women are unable to own land in much of the developing world and thus their access to land depends on the existence of communal lands or temporary cultivation rights granted by their husbands or families.[23] According to the FAO, there would be roughly 150 million fewer people living in hunger if women farmers' access to resources were equal to that of men's.[24] Further, women face food discrimination within their own households. Often women and girls eat only after the male members of the household have eaten.

Structural racial inequalities are prominent in every facet of contemporary food systems, from employment, to land ownership, to food access. People of color employed in food production, processing, distribution, retail, or service are largely concentrated in low-wage and dangerous jobs. For example, 65 percent of U.S. farm workers—who make, on average, $8.69 per hour—are Latino.[25] Blacks and Latinos comprise 62 percent of the meatpacking and poultry-processing workforce.[26] Meatpacking is an extremely dangerous job.

One in seven farmers in the United States was African American in 1920. By 2007, this number had dwindled to one in seventy.[27] USDA policies had much to do with the decline. A study done in 1964 by the U.S. Civil Rights Commission concluded that the USDA actively discriminated against African

American farmers. Loan agencies of the USDA systematically denied loans to black farmers and denied them disaster relief. Further, the study found that it was common practice to deny credit to anyone who participated in or assisted activities associated with the civil rights movement, who was a member of the NAACP, or who registered to vote or even signed a petition.[28]

People of color suffer from food insecurity at a much higher rate in the United States than do whites. Black and Hispanic households experience food insecurity at a rate that is substantially higher than the national average and more than twice that of white households. In fact, more than one out of four black households in the United States was food insecure at some point in 2013.[29] There are no national statistics on Native Americans, but a study done at the Pine Rest Reservation in South Dakota found that 40 percent of residents were food insecure. Another study, conducted in four southwestern Native American communities, found that 45 percent of adult residents and 29 percent of children were food insecure.[30]

Further, a study by Kelly M. Bower, of the Johns Hopkins School of Nursing, and colleagues from the Bloomberg School of Public Health demonstrated that neighborhoods in the United States that are predominantly African American suffer from the most limited access to venues that sell fresh, healthy food, *regardless of income*. In other words, when considering economically similar neighborhoods, those that are predominately black have more limited access to healthy food than those that are predominately white or Hispanic.[31]

These kinds of statistics are not restricted to the United States. Canadian Aboriginal peoples who do not live on a reservation are twice as likely to be food insecure as non-Aboriginal Canadians.[32] Forty percent of Roma children living in Central and Eastern Europe live in hunger.[33] Poverty—and its correlate of food insecurity—is the likely result for migrants to Europe, especially those who migrate from a non-EU nation. In Belgium, Luxembourg and Finland, as many as 40 percent of non-EU migrants are at risk of poverty. The risk of poverty is 30 percent or higher in nine of twenty-five EU countries, and is more than 20 percent in another nine countries.[34]

Globally, the highest rates of food insecurity occur in Africa, Latin America and Asia. All nations in this group are nonwhite and non-English speaking. Over half a billion people in Southern Asia live in chronic hunger. One in four people in sub-Saharan Africa is chronically hungry.[35]

These facts and statistics only scratch the surface of the ways in which structural racial inequalities are prominent in contemporary food systems. Much scholarship has been done, for example, on racial identity and food. It has been argued that the preparation and consumption of food are essential to racialized identities and belonging. Yet food histories of marginalized, nonwhite people are often "ignored, appropriated, or maligned" by those in power.[36]

Many food justice movements in Western, developed nations see local food as the solution to current injustices in our food system. However, locavorism in developed countries tends to be a movement embraced primarily by white, upper-middle-class individuals and tends to reflect white culture and values.[37] The nostalgic vision of farming so prevalent in narratives surrounding "local food" ignores much of the unjust history of farming practices in the West.[38] As Julie Guthman writes, speaking of the United States, "land was virtually given away to whites at the same time that reconstruction failed in the South, Native American lands were appropriated, Chinese and Japanese were precluded from landownership, and the Spanish-speaking *Californios* were disenfranchised of their ranches."[39] It should be clear why the nostalgia for agricultures of the past is not embraced by all. Local food narratives tend to erase the histories of those who were ostracized, marginalized and disenfranchised.

Further, given the way things currently stand, it is not clear that locally grown food is always the best choice if one is concerned about issues of global justice. In wealthy nations, those who eschew imported food may very well be harming peasant farmers in developing nations. Given current agricultural markets, such farmers depend on exports for their income. One could argue that food justice movements in developed nations are thus leaving out, albeit unintentionally, those who are wronged the most by the global food system.

OVERVIEW OF *JUST FOOD: PHILOSOPHY, JUSTICE AND FOOD*

Just Food: Philosophy, Justice and Food is divided into four main sections. The issues addressed in this volume are interdependent and intertwined, so the divisions are fluid and somewhat artificial. Nonetheless, four primary themes are discernible, and the volume is divided according to those themes. The four sections are: Food Access, Food Systems, Gender and Food and Local Food.

The chapters in part I of *Just Food: Philosophy, Justice and Food* offer philosophical analyses of injustices connected to food access. J. Michael Scoville sets the stage in chapter 1. He argues that articulating an account of sustainability is a prerequisite for answering more specific normative questions regarding food justice. Different accounts of sustainability will yield different results. Chapters 2 and 3 explore the notion of property and how it impacts food injustices. There is a rich philosophical literature on property and its justification. In chapter 2, Steven Minister utilizes this tradition by focusing on three philosophers from three different intellectual traditions: Thomas Aquinas (Christian theology), John Locke (modern liberalism) and Emmanuel Levinas (postmodernity). Minister argues that on all three accounts, property is justified because it satisfies human needs. As such, claims to property lose their justification when said claims deprive others of things they need. Minister argues that the human need for subsistence thus trumps property rights.

In chapter 3, J. M. Dieterle examines food deserts in light of the Lockean theory of property. Dieterle argues that any theory of justice that depends on or substantially includes a Lockean theory of property should admit that food deserts are a serious matter of justice. As such, limitations on property rights to correct the injustice are warranted. Jennifer Szende also addresses food deserts in chapter 4. She uses food deserts to illustrate Iris Marion Young's account of justice. Szende argues that food deserts are partially constituted by inequality of food distribution, but they are ultimately the product of oppression, domination and powerlessness.

The essays in part II examine the values that drive the debate over proper food systems. Should those values be efficiency and productivity? Or, instead, should democratic control of food systems be paramount? Chapter 5, by Ian Werkheiser, Shakara Tyler and Paul B. Thompson, discusses the food sovereignty movement. Werkheiser, Tyler and Thompson point to two different conceptions of food sovereignty. According to the first conception, which the authors call "participatory food sovereignty," the goal is *reform* of the food system. The second conception, "radical food sovereignty," calls for *resistance* to institutions that are dominant in the global food system and focuses on creating alternatives to those institutions.

In chapter 6, Mark Navin takes up a particular value embraced by the food sovereignty movement: gender justice. The end of gendered oppression and the ideal of gender equality are core components of food sovereignty, but one might wonder why that would be. What does gender justice have to do with ensuring that peasants have control over their food system? Navin attempts to clarify the relationship between gender justice and peasant self-determination.

Steve Tammelleo examines the consequences of the NAFTA, a specific example of neoliberal trade policies, in chapter 7. Tammelleo demonstrates the effect of liberalized trade on Mexican farmers and agricultural workers. In the end, he defends free trade agreements, with the caveat that they must be implemented in a fair and just manner. NAFTA was not, he argues, because the United States continues to subsidize its own agricultural products. As a result, special interest groups in the United States are benefited at the expense of agricultural workers in Mexico.

The essays in part III of *Just Food: Philosophy, Justice and Food* look at the ways in which gender impacts issues surrounding food. In chapter 8, Lori Watson examines the gendered inequities in the global food system. She focuses on three areas: food access, agricultural labor and water access. In each case, Watson demonstrates the disproportionate ways in which women and girls are impacted. She argues that global food security is a matter of gender justice and so requires gender equality and the empowerment of women.

Nancy M. Williams offers a Foucauldian analysis of masculinity and meat eating in chapter 9. Williams argues that although meat eating is conceptualized as a masculine activity, it is, instead, a disciplinary practice that produces subjected bodies and identities.

In chapter 10, Margaret Crouch examines the ways in which gender and food are portrayed in mainstream film. Each of us inhabits what Crouch calls a "food world." Our food worlds encompass what food and the activities surrounding food mean to us. Crouch demonstrates that, in film, the food worlds inhabited by men are substantially different than those inhabited by women. She argues that gender norms and heteronormativity dictate the relationships men and women have with food in mainstream film.

Part IV critiques the locavore movement.

In chapter 11, Nancy E. Snow examines what it might mean to be virtuous with respect to our food choices. She argues that the values of sustenance and sustainability are central to virtuous food shopping and eating. Given that food that is grown and marketed in ways consistent with these values is available, it initially seems possible to make virtuous food choices. However, many people in developed nations are not economically positioned to be able to make such choices. The larger context of systemic economic injustice renders putatively virtuous food choices as privileges had only by those able to afford them. Further, when we choose to avoid food produced by agribusiness and focus on local, sustainably grown food, we risk harming the livelihoods of those who are forced to participate in agribusiness just to make a living.

Chapter 12 offers another critique of the locavore movement. Liz Goodnick argues against locavorism insofar as it requires or permits one to eat locally raised meat. Nonhuman animals have inherent value, according to Goodnick, and eating them is immoral.

No single collection of essays can fully address the myriad inequities and injustices connected to food and contemporary food systems. For example, the injustices faced by those employed in food production, processing, distribution, retail and service are not directly addressed. More than half of the workers in the U.S. food system live below the poverty line. And, ironically, food insecurity is more prevalent in food system workers than in the general population.[40] Nor are the injustices and hardships faced by smallholder farmers in developed countries directly addressed in this volume. Many such farmers are under tremendous pressure to corporatize their farming practices and production methods.

Although issues of race, ethnicity and class are intertwined with the issues covered in this volume, and underpin many of the discussions in its pages, there is no essay specifically devoted to race and food. This is a significant omission.[41]

Another important omission involves food and disability. Recent research has demonstrated that disability significantly impacts a household's food security. Food insecurity is more common in households affected by disability, and it also tends to be more severe.[42]

Further, there are issues of power and control for those who require assistance or accommodation in eating. Who controls food choice? How does this affect one's independence?

Despite these omissions, I hope that *Just Food: Philosophy, Justice and Food* makes a significant contribution to philosophical scholarship on food and justice.

NOTES

1. Peter Singer, "Famine, Affluence, and Morality," *Philosophy and Public Affairs* 1 (1972): 229–43.

2. See, for example, James P. Sterba, "The Welfare Rights of Distant Peoples and Future Generations: Moral Side Constraints on Social Policy," *Social Theory and Practice* 7 (1981): 99–119; Onora O'Neill, *Faces of Hunger: An Essay on Poverty, Justice and Development* (Unwin Hyman, 1986); Thomas Pogge, "'Assisting' the Global Poor," in *The Ethics of Assistance: Morality and the Distant Needy*, ed. Deen K. Chatterjee (Cambridge: Cambridge University Press, 2004), 260–88; Hugh LaFollette, *World Hunger* (Malden, MA: Blackwell, 2003) and *World Hunger and Morality*, ed. William Aiken (Englewood Cliffs: Prentice Hall, 1996).

3. United Nations Universal Declaration of Human Rights, Article 25 (1): "Everyone has the right to a standard of living adequate for the health and well-being of himself and of his family, including food, clothing, housing and medical care and necessary social services, and the right to security in the event of unemployment, sickness, disability, widowhood, old age or other lack of livelihood in circumstances beyond his control." http://www.un.org/en/documents/udhr/.

4. World Food Programme, "Who Are the Hungry?" accessed March 25, 2015, http://www.wfp.org/hunger/who-are.

5. Alisha Coleman-Jensen, William McFall, and Mark Nord, "Food Insecurity in Households with Children: Prevalence, Severity, and Household Characteristics, 2010–11," USDA Report Summary, 2013.

6. FAO, "Trade Reforms and Food Security: Conceptualizing the Linkages," Rome: Food and Agricultural Organization of the United Nations (2003): 29.

7. World Health Organization (WHO), "Food Security," accessed July 15, 2014, http://www.who.int/trade/glossary/story028/en/. The FAO adds a fourth: stability (that access to food is stable over time and not subject to shocks, economic or otherwise). The WHO criteria fold stability into facet (1) by including "on a consistent basis." See FAO, "Food Security," Rome: Food and Agricultural Organization of the United Nations Policy Brief 2 (2006).

8. FAO, "World Food Summit: Food For All," Rome: Food and Agricultural Organization of the United Nations (1996).

9. This figure includes only waste at the retail and consumer levels. See Jean C. Buzby, Honan Farah Wells, and Jeffrey Hyman, "The Estimated Amount, Value, and Calories of Postharvest Food Losses at the Retail and Consumer Levels in the United States," USDA Report, 2014.

10. Love Food Hate Waste, "The Facts About Food Waste," accessed April 20, 2015, http://england.lovefoodhatewaste.com/node/2472.

11. European Commission, "Food Waste," last updated March 9, 2015, http://ec.europa.eu/food/safety/food_waste/index_en.htm.

12. UN News Centre, "World Population Projected to Reach 9.6 Billion by 2050—UN Report," accessed July 7, 2014, http://www.un.org/apps/news/story.asp?NewsID=45165#.U78ySvk7uSp.

13. FAO, "Global Agriculture Towards 2050," Rome: Food and Agricultural Organization of the United Nations (2009).

14. C. Ford Runge and Benjamin Senauer, "A Removable Feast," in *The Ethics of Food: A Reader for the 21st Century*, ed. Gregory E. Pence (New York: Rowman and Littlefield, 2002), 182.

15. Hannah Wittman, Annette Desmarais, and Nettie Wiebe, "The Origins and Potential of Food Sovereignty," in *Food Sovereignty: Reconnecting Food, Nature, and Community*, ed. Wittman, Desmarais, and Wiebe (Oakland, CA: Food First, 2010), 3.

16. Marceline White, Sarah Gammage, and Carlos Salas Paez, "NAFTA and the FTAA: Impact on Mexico's Agricultural Sector," IATP Institute for Agriculture and Trade Policy, 2003, accessed July 10, 2014, http://www.iatp.org/files/NAFTA_and_the_FTAA_Impact_on_Mexicos_Agricultu.pdf.

17. For extensive discussion, see Paul Altidor, *Impacts of Trade Liberalization Policies on Rice Production in Haiti* (MIT Master's Thesis, 2004).

18. Declaration of Nyéléni, adopted at the Forum for Food Sovereignty, Sélingué, Mali, February 27, 2007. Full text available here: http://www.nyeleni.org/spip.php?article290, accessed June 14, 2014.

19. The locution "fringe food locations" originates with the Mari Gallagher Research and Consulting Group. For their research on food deserts, see http://marigallagher.com/.

20. See Mari Gallagher, "Food Desert and Food Balance Community Fact Sheet," Mari Gallagher Research and Consulting Group (2010), accessed June 1, 2013, http://www.fooddesert.net/wp-content/themes/cleanr/images/FoodDesertFactSheet-revised.pdf.

21. See Gallagher, "Fact Sheet"; "Examining the Impact of Food Deserts on Public Health in Detroit," Mari Gallagher Research Group (2007); "Examining the Impact of

Food Deserts on Public Health in Chicago," Mari Gallagher Research and Consulting Group (2006); and other publications at http://marigallagher.com/.

22 See FAO, "Gender Equality and Food Security," Rome: Food and Agricultural Organization of the United Nations (2013), accessed June 19, 2014, http://www.fao.org/wairdocs/ar259e/ar259e.pdf.

23. See, for example, Bina Agarwal, *A Field of One's Own: Gender and Land Rights in South Asia* (Cambridge: Cambridge University Press, 1994).

24. FAO, "10 Hunger Facts for 2014," Rome: Food and Agricultural Organization of the United Nations, (2013), accessed June 10, 2014, https://www.wfp.org/stories/10-hunger-facts-2014.

25. See Race Forward: The Center for Racial Justice Innovation, "Food Justice," accessed May 4, 2015, https://www.raceforward.org/research/reports/food-justice. For wage information, see the United States Department of Labor, "Occupational Employment Statistics," accessed May 4, 2015, http://www.bls.gov/oes/current/oes452099.htm.

26. Editors, "From the Editors," *Race/Ethnicity* 5 (2011): vii.

27. Other Worlds, "Uprooting Racism in the Food System: African Americans Organize," accessed May 1, 2015, http://otherworldsarepossible.org/uprooting-racism-food-system-african-americans-organize.

28. PBS, *Black Farming History: The Civil Rights Years (1954–1968)*, accessed May 4, 2015, http://www.pbs.org/itvs/homecoming/history5.html.

29. Alisha Coleman-Jensen, Christian Gregory, and Anita Singh, "Household Food Security in the United States in 2013," USDA Economic Report, September 2014, Table 2. 26.1 percent of black households and 23.7 percent of Hispanic households in the United States suffer from food insecurity, compared to 10.6 percent of whites. The national average is 14.3 percent.

30. Katherine W. Bauer, Rachel Widome, John H. Himes, Mary Smyth, Bonnie Holy Rock, et al., "High Food Insecurity and Its Correlates Among Families Living on a Rural American Indian Reservation," *American Journal of Public Health* 102, no. 7 (2012): 1346–52; and Brita Mullany, Nicole Neault, Danielle Tsingine, Julia Powers, Ventura Lovato, et al., "Food Insecurity and Household Eating Patterns Among Vulnerable American-Indian Families: Associations with Caregiver and Food Consumption Characteristics," *Public Health Nutrition* 16, no. 4 (2013): 752–60.

31. Kelly M. Bower, Roland J. Thorpe Jr., Charles Rohde, and Darrell J. Gaskin, "The Intersection of Neighborhood Racial Segregation, Poverty, and Urbanicity and its Impact on Food Store Availability in the United States," *Preventative Medicine* 58 (2014): 33–9.

32. Council of Canadian Academics, "Aboriginal Food Security in Northern Canada: An Assessment of the State of Knowledge," accessed May 4, 2015, http://www.scienceadvice.ca/en/assessments/completed/food-security.aspx.

33. Henry Scicluna, "The Health Situation of the Roma in Europe," 4th Conference on Migrant and Ethnic Minority Health in Europe, June 21–23, 2013, Universita Bocconi, Milan, Italy.

34. Orsolya Lelkes and Eszter Zólyomi, "Poverty and Social Exclusion of Migrants in the European Union," European Centre Policy Brief (March 2011).

35 FAO, "Food Security Statistics," accessed May 7, 2015, http://www.fao.org/economic/ess/ess-fs/en/.

36. For a summary and discussion of this scholarship, see Rachel Slocum, "Race in the Study of Food," *Progress in Human Geography* 35 (2011): 303–27. Quotation is from 307.

37. See Julie Guthman, "'If Only They Knew': The Unbearable Whiteness of Alternative Food," in *Cultivating Food Justice: Race, Class, and Sustainability*, ed. Alison Hope Alkon and Julian Agyeman (Cambridge, MA: MIT Press, 2011), 263–81. Guthman argues that many of the idioms of alternative food practice "are either insensitive or ignorant (or both) of the ways in which they reflect whitened cultural histories and practices. 'Getting your hands dirty in the soil,' 'if only they knew [where their food came from],' and 'looking the farmer in the eye' all point to an agrarian past that is far more easily romanticized by whites than others," 275.

38. See Guthman, "If Only They Knew." See also Slocum, "Race in the Study of Food."

39. Guthman, "If Only They Knew," 276.

40. Joann Lo, "Racism, Gender Discrimination, and Food Chain Workers in the United States," in *Global Food Systems: The Issues and Solutions*, ed. William D. Schanbacher (ABC-CLIO, 2014), 60.

41 Alkon and Agyeman, *Cultivating Food Justice*, is an excellent collection of essays by social scientists on the connections between race and injustices related to food.

42. Alisha Coleman-Jensen and Mark Nord, "Disability is an Important Risk Factor for Food Insecurity," USDA Economic Research Service (2013), accessed March 25, 2015, http://www.ers.usda.gov/amber-waves/2013-may/disability-is-an-important-risk-factor-for-food-insecurity.aspx#.VRLO946jOSo.

Part I

FOOD ACCESS

Chapter 1

Framing Food Justice

J. Michael Scoville

The discussion of food justice norms tends to be focused on three main concerns.[1] One is distributive issues, for example, whether all people have access to safe and healthy food, or whether everyone working within the food system is paid fairly and able to work in a safe environment. A second concern is issues of representation and political voice. Here the focus is whether all people are capable of participating in relevant decision making and the construction of public policies relating to the production, consumption and distribution of food. A third concern is the normatively significant connections between, on the one hand, the values of food and food-related practices, and on the other, collective self-determination. This third focus is often expressed in terms of food sovereignty, and has its origin in peasant social movements, notably, La Via Campesina. Discussion of these three concerns is complex, subject to ongoing disagreement and practically fraught.

Different views of what justice requires always reflect particular framings of what questions or concerns are thought to be crucial. Not surprisingly, the question of which framing (or framings) is best is controversial. With this in mind, I assume it is critically important to consider sustainability as a relevant framing for any contemporary theorizing about justice. Articulating an account of food justice, specifically, in isolation from broader questions about sustainability would leave many important normative issues unaddressed. The primary aim of this chapter is to explore how our thinking about food justice norms might be guided, constrained and in general enriched if we consider these norms in relation to sustainability.

A difficulty for this proposed focus is that many philosophers (among others) have viewed the concept of sustainability with suspicion. Reasons for this range from concern about sustainability being hopelessly vague and hence useless for policy, to concern that interest in sustainability is just the latest cover for business as usual and thus a betrayal of the environmental cause. While I believe such concerns are unconvincing, there is no question that sustainability is a contested concept—one that needs careful specification and defense if it is to do any work helping to frame discussions of food justice.

I assume that a significant reason to care about sustainability is the worry that we are shortchanging future generations through our collective conduct, giving them less than is their due.[2] This is partly a matter of justice, but it is also a broader question of what we ought to be doing to preserve conditions that will make life worth living in the future.[3] With this in mind, a fundamental aim of discussions of sustainability should be to clarify the X that we ought to be preserving, insofar as we can, for future generations both as a matter of justice and as a condition for living worthwhile lives.[4] The challenge is to clarify the relevant X and the normative account that supports it.

In my reading of the literature, there are basically three types of sustainability views. One, which I'll call "the minimalist view," aims to specify our obligations to present and future generations (of human beings) in terms of maintaining the capacity to be well off. A second, which I'll refer to as "the human flourishing account," rests on the belief that human beings need access to a variety of specific and disaggregated goods, experiences and relationships in order to achieve well-being.[5] A third incorporates aspects of the first two, but includes in addition nonanthropocentric reasons; I'll call this "the demanding view." Depending on the view of sustainability one adopts, there can be significantly different implications for how we should think about, and try to realize in practice, food justice. I explore some of these implications with respect to each type of sustainability view sketched.

I. THE MINIMALIST VIEW

A number of economists and philosophers defend something like what I'm calling the minimalist view.[6] Despite differences in detail, defenders of this view more or less share three basic commitments. First, the core ethical commitment of the minimalist view is that the obligation of sustainability requires the present generation to aim at enabling all people, present and future, to have the option or capacity to be well off.[7] With respect to the X that we ought to be sustaining, defenders of the minimalist view answer that our collective aim should be to maintain a nondeclining stock of total capital assets, which is assumed to be necessary for maintaining welfare over time. This stock is understood broadly and includes a diversity of things—for example, infrastructure, knowledge, technology and savings and investment, as well as the various resources and life-support functions provided by nature (commonly referred to as "natural capital").[8] Of course, there could be a nondeclining stock of the relevant goods and yet people might lack access to it. So defenders of the minimalist view should be read as assuming that all people should have access to the relevant goods as a matter of basic justice.

A second commitment of the minimalist view concerns the conception of welfare or well-being that is presupposed. Some prominent defenders of this view assume a desire or preference satisfaction account, where welfare consists in the satisfaction of an individual's desires or preferences.[9] Such a view faces serious difficulties. Desires and preferences are highly adaptable, largely dependent on what is, or is expected to be, available, and can be distorted in ways that give us no reason to aim at satisfying them.[10] Further, the fact that desires and preferences are subject to great variation, dependent as they are on changing circumstances, presents a problem when we try to clarify the content of our obligations to future generations. After all, how can we know with any certainty what future people will desire or prefer? If we endorse a modest "ought implies can" principle, then a desire or preference satisfaction account has the result of potentially undermining, or at least leaving largely unspecifiable, our obligations to future people. This implication may be unintended, but that hardly removes the problem.

To avoid these difficulties, the minimalist view does best to incorporate a need-based conception of well-being. Though the specification of the relevant needs is theory dependent, and not without controversy, it seems reasonable to think that theorists and policy makers could clarify a set of "core" or basic needs that would focus and guide social and political decision making. A statement from James Sterba suggests the general idea here: "Basic needs, if not satisfied, lead to significant lacks and deficiencies with respect to a standard of mental and physical well-being. Thus, a person's needs for food, shelter, medical care, protection, companionship and self-development are, at least in part, needs of this sort."[11]

A third aspect of the minimalist view is a commitment to a permissive view of substitutability. The economist Robert Solow gives expression to this idea when he writes:

> Goods and services can be substituted for one another. If you don't eat one species of fish, you can eat another species of fish. Resources are, to use a favorite word of economists, fungible in a certain sense. They can take the place of each other. That is extremely important because it suggests that we do not owe to the future any particular thing. There is no specific object that the goal of sustainability, the obligation of sustainability, requires us to leave untouched.[12]

Clearly, assumptions about substitutability have a direct bearing on the question of whether we ought to be preserving some particular X in order to fulfill our sustainability-related obligations. While defenders of the minimalist view are not committed to the unlimited substitutability of (natural) goods and resources in practice, they are not opposed to this idea in principle.[13] This commitment makes the view blind to some important normative considerations. The human flourishing account helps to illuminate these considerations, and I'll turn to this now.

II. THE HUMAN FLOURISHING ACCOUNT

As a focus for my discussion of the human flourishing account, I'll consider the recent work of John O'Neill, Alan Holland and Andrew Light.[14] By articulating a normative basis for objecting to certain sorts of substitution, even if such substitutions are technically possible, the human flourishing account illuminates a significant potential shortcoming of the minimalist view.

The account developed by O'Neill et al. has two main elements. First, the authors defend a version of an objective state theory of well-being.[15] The usual list of objective states or goods is endorsed (physical health, personal relations, autonomy, etc.), with one notable addition: the good of having a well-constituted relation with the nonhuman world.[16] The relevant states are conceived as necessary constituents of a flourishing life, such that one is harmed if one lacks access to these states.[17] Further, the authors suggest that the goods in question are disaggregated, meaning, a lot of one good cannot substitute for too little or none of another.

The second aspect of the account rests on an appeal to the importance of historical considerations for our thinking about value, human well-being and the natural (or partly natural) world. To value something in a historical way is to value it in virtue of its particular history, or because it is the product of processes of a certain sort.[18] The contrast to a historical view is one that regards the value of an object as consisting solely in terms of its specific cluster of properties, where this cluster is understood in isolation from the history of the object or the processes by which it came about. O'Neill et al. focus on a subset of possible historical considerations characterized in terms of "narrative." Narrative considerations illuminate the ways in which particular environments are valuable because they embody the labors and history of individuals and communities.[19] On this account, being able to perform one's identity in, or in relation to, narratively significant environments is partly constitutive of living a flourishing life. Put differently, one's life can be intimately bound up with a place in such a way that physical continuity with that place helps to "make sense" of one's life; conversely, alienation from a narratively significant place can diminish one's life.[20] According to O'Neill et al., an objective state theory of well-being must be tempered by narrative considerations if we are to adequately appraise how a person's life goes.[21]

The account offered by O'Neill et al. has clear implications for debates about sustainability. The authors argue that "what we need to pass on to future generations is a bundle of goods that can maintain welfare across the different dimensions of human life."[22] These goods include the objective state components noted above, as well as the various life-support functions of the natural world that figure prominently in the minimalist view (i.e., as a precondition for satisfying basic needs). There are, in addition, two main respects in which O'Neill et al. offer something that goes beyond the minimalist view. First, the authors emphasize the social and environmental context in which human flourishing occurs. This emphasis is especially clear in the focus on narrative significance. Second, O'Neill et al. claim that we are obligated to maintain "particular environments" in order to fulfill our sustainability-related obligations. Both of these aspects have clear implications for how we should think about normatively permissible substitutions, some of which have a direct bearing on issues related to food justice. I will consider each of these aspects in turn.

Defenders of the minimalist view typically say very little about the social and environmental context in which welfare needs arise and are satisfied. To discuss well-being in abstraction from the broader context of supportive relationships, social practices and specific environments presents some notable hazards. Recall Solow's remark regarding the substitution of one species of fish for another. By endorsing a permissive view of substitutability, defenders of the minimalist account deny or obscure a number of normatively significant issues. To stay with the fish example, some communities around the globe are dependent on fish as an important source of food. Fish and fishing are also intimately tied up with their way of life. Due in large part to the shift toward industrial fishing and aquaculture since the 1970s, aquatic ecosystems have become degraded to the point that in some places traditional fishing communities can no longer catch the fish they need.[23] To think, as Solow urges us to, that the main issue here is compensating people so they can find a suitable food substitute for fish is to miss a number of specific harms and injustices experienced by the people in question. Not only are the communities losing their access to an important food source, they are also, more generally, losing their ability to self-provision from nature in order to meet their needs. They may also be experiencing the destruction of their way of life. All of these losses may be very significant harms to the people involved. There may also be serious injustices, for example, if the people in question are (or have been) unfairly disadvantaged by a pattern of industrial development that they had little or no ability to influence the shape of.[24] There is much more to say about such cases, but the main point is that Solow's claim "If you don't eat one species of fish, you can eat another species of fish" is naïve and very likely to lend support to policies that would generate, and in some cases further exacerbate, significant harm and injustice.

Perhaps a defender of a Solow-type view could say, in reply, that the minimalist view can account for these harms and injustices. That is, if the replacement of one species of fish with another goes hand in hand with the demise of traditional fishing practices, and if the demise of these practices makes the people in question worse off, then the substitution has resulted in a diminution of welfare. The problem is this reply is compatible with, and does not challenge, the view that the loss in question is one for which there could be adequate compensation. But this assumption about compensation is implausible. A way of life is not merely a way of meeting one's needs, or maintaining an abstract set of options for oneself and one's descendants. It involves a lived relation to particular environments and objects (such as fish). The relevant environments and objects are often suffused with meaning and cultural significance. To think that the people in question are—or in principle could be—adequately compensated so long as a substitute is found for fish, fishing and the specific environments (or environmental amenities) at issue, is to miss the specific type of harm and injustice caused, with its attendant losses. However, if the relevant individuals and communities actually regarded the compensation as acceptable, then this rejoinder would be considerably weakened.[25] Though even in that case, we might still inquire whether people are accepting the compensation because they really have no alternative. There is, in any case, a whole set of considerations here—about the nature of harm, about injustice, about power—that the defender of a permissive view of substitutability obscures, or is apt to obscure.

These considerations pose a general problem for any endorsement of a permissive view of substitutability. Let me note three further points that emphasize the importance of the preceding discussion for our thinking about sustainability. First, harm and injustice, such as that caused by industrial aquaculture, is often a driver of environmental degradation. Second, I assume we need models or paradigms of ways of life that are sustainable. Indeed, one might think that an important aspect of our obligation to future generations is to provide them with models of how to live sustainably. When industrial aquaculture displaces traditional aquaculture (to stay with this case, though the point is a general one), this often destroys sustainable ways of life. This is bad, first and foremost, for those whose way of life is undermined.[26] But it is also bad in the sense that we lose a paradigm from which people might learn how to live in particular places, meeting their needs and creating culture, while not destroying biodiversity, soil fertility and so on. More is at stake here than one might initially think.

A third point merits special emphasis. There is a common feature of philosophical debates about distributive injustice, particularly at the global level. Authors employ categories such as "the global rich" and "the global poor," and debates focus on what those who are comparatively well off owe those less well off, or at least those considered "poor" or "least well off." These debates typically assume an account of well-being that emphasizes certain absolute, rather than relative, dimensions in order to designate who is poor or badly off. Lacking well-being is thus partly understood in terms of being unable to consume adequate calories or protein, being unable to access clean air and water and so on.

As stated, there is nothing particularly objectionable here. The problem arises when those designated poor are considered so according to certain measures, such as having the power to purchase modern consumer goods, industrial agricultural technologies, and the like, while these same people are decidedly *not* poor in other important respects. For example, many peasant and subsistence communities around the globe are capable of self-provisioning from their local environments. They have enough to eat, are relatively healthy, have access to clean water, and so on. This is important because it means that, at least in these cases, people designated as poor are not necessarily poor in a sense that justifies a poverty-removal project. Poverty-removal projects are problematic if they assume a questionable notion of human welfare, such as being able to participate in modern consumer society, or being able to purchase the goods and services of industrial agriculture. Defenders of such projects too often fail to recognize the projects' negative effects on the lives of those whom they putatively aim to help—not to mention the considerable negative effects of such projects with respect to environmental quality.[27] Here it is useful to distinguish between poverty as subsistence and poverty as (absolute) deprivation.[28] Only the latter may justify a poverty-removal project.

The point is not to romanticize subsistence living (though its merits are likely underappreciated), but to recognize that subsistence living is often much better than the situation people find themselves in when they are forced off their lands and are no longer able to self-provision from nature. Importantly, subsistence ways of life solve a social challenge (people being able to meet their needs in culturally appropriate ways), while also preserving what are often ecologically and socially sustainable ways of life.[29] Obviously, we have good reasons to support the preservation of subsistence communities insofar as they represent so-called win-win cases.

A related point merits mention here. The relationships and practices that exist in, or that might be created by, a particular community in order to meet individual and communal needs can be both supportive and expressive of values and commitments that are themselves valuable, that is, beyond the specific good of having the relevant needs satisfied. For example, self-provisioning from nature, or satisfying individual and communal needs in ways that allow for collective self-determination and (cultural) self-expression, can provide a corrective to legacies of colonialism or ongoing social and political marginalization. These legacies, with all of their accompanying harms and wrongs, risk being further entrenched by views that conceptualize human needs, and their satisfaction, in abstraction from social, environmental and political contexts. Further, the idea that state institutions or various non-governmental organizations could provide the relevant need satisfaction—a common assumption among theorists who focus on distributive justice with regard to food—misses the normative significance of people being able to meet their needs *themselves*, and in ways that express communal and cultural values.[30] The human flourishing account can readily appreciate this point, in contrast to the minimalist view that tends to focus almost exclusively on distributive concerns.

To illustrate, consider two of O'Neill et al.'s objective goods: being adequately nourished (an aspect of physical health[31]) and being socially affiliated. The normative appeal of O'Neill et al.'s kind of view does not lie merely in the fact that it highlights these goods *as goods*. That is something the minimalist view can also appreciate (insofar as these goods are understood as, or as related to, basic needs). Rather, its appeal concerns the way these goods are viewed with reference to supportive contexts, relationships and practices. Being adequately nourished may be intimately tied to being socially affiliated in certain ways. Importantly, both goods are made possible and supported by particular environments and social practices. It follows that one cannot access the good in question without also accessing the relevant environments and participating in the appropriate social practices.[32]

While this point has general significance, it seems particularly salient for thinking about collective self-determination in the case of peasant and indigenous communities around the world. Although collective self-determination is not a good explicitly emphasized by O'Neill et al., I think the human flourishing account has theoretical resources for appreciating this good. If, as Kyle Whyte suggests, "collective self-determination refers to a group's ability to provide the cultural, social, economic and political relations needed for its members to pursue good lives,"[33] then the human flourishing account appears already committed to this idea.

The preceding discussion brings into focus a second aspect of the human flourishing view, one that further suggests its possible advantage over more minimalist accounts. I'm referring here to our obligation, argued for by O'Neill et al., to maintain particular environments as a matter of sustainability. The crucial issue, of course, is clarifying the content of this idea.

The authors discuss two different senses of the relevant particularity, though I'll only consider one here.[34] This sense refers to those environments that are (partly) constitutive of communities, their values and their collective self-determination.[35] The reasoning here is that ensuring access to specific goods, such as being adequately nourished and socially affiliated in ways conducive to flourishing, requires that we maintain the cultural and physical conditions for certain kinds of community and social practice. Examples relevant to food justice are numerous, but I'll borrow one from Kyle Whyte that is revealing. For the indigenous Anishinaabek in the Great Lakes region, *manoomin* (wild rice) has a special, hub-like status. Such foods bring together, and suggest the deep connections between, many different aspects of a community or culture's way of life—aspects that are biological, ecological, cultural, economic, spiritual and so on. "Access to the nutritional value of manoomin," writes Whyte, "requires family, economic, social and political relations; these relations are, in turn, made possible through manoomin. Other foods, such as the commodity cheese and spam distributed to some Anishinaabek through U.S. food assistance programs, or microwave meals, cannot replace manoomin as comparable contributors to the establishment and maintenance of these relationships."[36] If well-intentioned food assistance programs tried to provide culturally appropriate food, for example, by making available packaged "wild rice," this would provide relevant nutritional value. But this would obviously fail to provide a substitute for *manoomin*, given the embeddedness of the latter in a set of significant food-related practices (seasonal group activities of gathering, processing, etc.) and particular environments (the shallow, clear, slow-moving waterways suitable for the rice to grow). The human flourishing account provides a normative basis for understanding the significance and nonsubstitutability of certain goods or aspects of the world, such as *manoomin* and the cultural and ecological contexts that support its flourishing.

In concluding this section, I'll note one difficulty for the human flourishing view and highlight two possible virtues of the account. First, the difficulty. If one includes within the objective account of human well-being such goods as autonomy, as O'Neill et al. do, then it seems that respect for autonomy will likely generate serious conflicts over exactly what the constituents of well-being are. After all, respecting others' autonomy would surely include respecting others' freedom to make up their own minds about what their own good consists in. The account thus appears threatened with foundering on the ground of reasonable disagreement concerning what human flourishing consists in.[37] More needs to be said to address this issue.

Despite this difficulty, the human flourishing account has at least two possible virtues. First, it presents a number of considerations that would block, or at least greatly complicate discussions of, acceptable substitutions with respect to the natural (or partly natural) world. For example, if appreciating certain environments (such as culturally significant landscapes) or other species (such as *manoomin*) is relevant to human well-being in ways that matter to particular communities, then this would provide at least a strong *prima facie* reason against harming these environments or species, or replacing them with something else. In contrast, the minimalist view seems unable to appreciate how particular goods (e.g., culturally significant environments or species) might be (partly) constitutive of human flourishing, such that harming these goods, or substituting them with other goods, may involve a significant loss in welfare. The relevance of this claim for debates about food justice has already been emphasized: I assume there are specific foods, and food-related practices (e.g., fishing, gathering, etc.), that promote individual and collective well-being in ways that would be seriously weakened or undermined were these foods made unavailable, or the food-related practices made impossible to perform.

A second possible virtue of the human flourishing account is that it is a form of anthropocentrism, albeit a rich and complex one. Perhaps this feature of the view increases its chances of engaging public interest, motivating action and influencing public policy. Pragmatic considerations aside, it is noteworthy that the human flourishing view does not simply regard human well-being as decisive with regard to ethical decision making, at least not in any simple way. This might provide some solace to those who believe, as I'm inclined to, that nonhuman nature makes a claim on us independently of human interests. If the human flourishing account is concerned to articulate, among other things, attitudes that we have reason to cultivate in ourselves so as to flourish, then it can presumably incorporate the idea that we ought to overcome within ourselves the attitude that the natural world is simply there *for us*. It is not enough that our use of the natural world is fair, our distributions just and that all people are empowered to participate in decision making about how the world is used. More is at stake in living a worthwhile life than that. It is of interest that one can reach this conclusion without departing from anthropocentric commitments.

III. THE DEMANDING VIEW

To appreciate the distinctiveness of the third view of sustainability, what I'm calling "the demanding view," I need to introduce two concepts from the scientific and environmental ethics literature. The first is that of ecological health, a technical and explicitly normative concept. I understand ecological health to refer primarily to two properties of natural (or partly natural) systems: (1) the counteractive capacity to withstand stress or change (often glossed in terms of "resilience") and (2) the capacity of natural systems to function well over the long term, thus providing a range of ecosystem services (e.g., nutrient cycling, soil production, waste assimilation, etc.).[38] The second concept is biological or ecological integrity (or "integrity" for short). Integrity refers to a property of landscapes that are relatively unmodified by human activity and that have their native biota largely intact.[39] By "native biota," I mean the native plant and animal life in a given place, and whatever ecological relationships these instantiate.

A central commitment of advocates of the demanding view is that we ought, in our collective action, to aim at preserving or restoring both ecological health and integrity. It is, arguably, concern for integrity specifically that makes the demanding view *demanding*, given that (for reasons I'll explain shortly) concern for ecological health is already implied by the minimalist and human flourishing views. Of course, how demanding the third view is will depend on how much integrity we ought to be preserving or restoring (more on this later).

There is a possible connection between ecological health and integrity that is worth noting. Landscapes with their integrity intact are commonly thought to be instrumentally important to areas that exhibit, or might come to exhibit, ecological health.[40] The basic idea is that integrity areas are a storehouse of resources that ecologically healthy areas might need in order to be replenished and kept vital over time. If ecological health is viewed as valuable because it supports human well-being—something defenders of both the minimalist and the human flourishing accounts (should) agree on—then integrity could also be viewed as instrumentally valuable to human well-being. A food-related example of this line of argument is the importance of wild biodiversity (exemplified by many, if not all, integrity areas) for agricultural biodiversity, which in turn is important for maintaining soil fertility, crop resilience and much else. Further, if certain culturally significant foods, such as *manoomin*, depend on a high degree of wild (or relatively wild) biodiversity, as well as relevantly intact natural ecosystems, then concern for integrity, at least in some areas and to some extent, would be appropriate for a defender of the human flourishing account.

One difficulty for this instrumental defense of integrity is that it might be empirically questionable whether sites with integrity are *in all cases* instrumentally important to maintaining ecological health, where the latter is understood as instrumentally important to human well-being. There seems to be nothing inherent in the idea of integrity to suggest that a state in which integrity obtains would *necessarily* conduce, whether directly or indirectly, to human well-being.[41] A second problem with this line of defense is that it reduces the ideal of integrity to something practically necessary, and thereby pushes to the background noninstrumental (as well as nonanthropocentric) reasons for caring about integrity.

In light of these concerns, a defender of the third view has reason to articulate a noninstrumental defense of integrity. The attraction of such a defense is twofold. First, it allows one to integrate an array of considerations that seem independently important and that might not be appreciated as being deeply connected. Clarifying the concept and value of integrity helps to make these connections clear. Second, the noninstrumental defense brings into focus certain nonanthropocentric reasons, and in that respect goes beyond both the minimalist and the human flourishing views. In what follows, I briefly comment on the noninstrumental reasons that seem crucial to explaining the value of integrity. Reflection on these reasons helps to reveal more fully the content of the demanding view. I conclude with some challenges that this type of view presents for thinking about food justice specifically.

One group of noninstrumental considerations is broadly aesthetic. Landscapes that exemplify a high degree of integrity would likely exhibit a number of properties that merit and sustain an aesthetic response, such as intricacy, multifaceted complexity and uniqueness. Further, there is a way of connecting the defense of integrity to the possible significance of nature's *otherness*, understood to mean nature that is largely the product of processes that do not embody human designs, purposes, or aspirations.[42] A world with a nontrivial amount of integrity is one that evinces considerable nonhuman otherness. If such otherness is normatively interesting, then that would be a reason to care about integrity.

A second set of considerations relates to the value of flourishing. Integrity consists partly in the presence of various species of plants and animals living in suitable ecological contexts. Maintaining the existence of these forms of life in the wild—without regard to their possible usefulness to us—is a central focus of defenders of integrity. In this respect, integrity gives expression to the idea of a variety of other forms of life flourishing in their own way. The fact that these forms of life have a good that is not necessarily our good, and that may even be at odds with our good, is something the defender of integrity recognizes and views positively.

This last point might be particularly persuasive when we consider the case of sentient beings. For all sentient animals, there is something it is like to *be* the animal in question. This means that sentient animals can care (in some meaningful sense) about what happens to them, regardless of whether or not anyone else does. A number of philosophers argue that this fact generates a reason for ethically sensitive beings like us to be concerned with the lives and goods of sentient animals. This reasoning applies to both wild and domesticated animals, but the case of wild animals is most relevant to the defense of integrity.

I'll note one final consideration here. Integrity gives expression to the thought that we are part of a living totality that has immense value.

This totality includes all of the elements that have been discussed: great complexity and uniqueness both at the level of forms of life, and of ecological wholes; the idea of forms of life that have a good that is not necessarily our good; and the thought that some of the forms of life in question are experiencing subjects, which raises the stakes of concern. These considerations suggest that respect for nature's integrity should be an important part of the goals that comprise sustainability. It follows that the preservation or restoration of significant portions of the world to a state of integrity should be at least a long-term goal for collective action. Importantly, the defense of integrity gives clear expression to the idea that nature makes a claim on us beyond the call of human needs and interests. What this means in practice is that we should constrain the pursuit of our good (however understood) out of respect for nature's integrity.

A number of challenges remain for the defense of the demanding view. I'll limit myself to addressing one issue: the relative importance of ecological health (or "health" for short) and integrity. At issue is the weight of the reasons we have to aim at promoting or respecting health and integrity as goals for collective action. Here is one way to think about this issue. Maintaining (and, as necessary, restoring) the health of those parts of the world that we have to inhabit and use to meet our needs should never be traded off against any other goal, economic or otherwise. The goal of maintaining health (at a nontrivial scale) should thus provide a fundamental constraint on how we inhabit and use the world. Maybe in some imaginable emergency situations, say of urgent socioeconomic hardship, ecological health can be sacrificed in some places, to some degree, and over the relatively short term.[43] Preserving ecological health is, or at least ought to be, a matter of prudential collective concern in the present. It is also a basis for securing intra- and intergenerational distributive justice (assuming that maintaining the capability of human beings to meet their needs from nature is a crucial aim of any plausible view of what distributive justice requires).

With regard to integrity, one could construe respect for this as an absolutist constraint on how we use nature, or as an important but defeasible constraint. (And, of course, one could opt for something even weaker.) Clearly, an absolutist constraint would be very demanding, and in the minds of many, implausible on that count. Understood as a defeasible constraint, respect for integrity is far less demanding, but still holds on to the core commitments of the demanding view. I won't try to settle this issue here. But I will note that a significant barrier to our taking respect for integrity seriously, whether as an absolutist or defeasible constraint, is that the acknowledgment of such a constraint in practice would likely entail substantial economic losses, or foregone development opportunities, for certain people and perhaps entire nations or groups of nations. This possibility raises difficult questions of justice. Indeed, the question of justice here is magnified in those cases where the people or nations that incur a loss, or are expected to forego a development opportunity, are currently impoverished and in need of meaningful development. Things are complicated by the fact that a high degree of integrity is currently found in many parts of the world that, not coincidentally, are also socioeconomically impoverished. Further, even if one thought respect for integrity in general was a very worthwhile aim, there might be reasons to prioritize maintaining areas of integrity that exemplify a high degree of biodiversity (or other especially valuable properties). This makes the matter at hand even more urgent. For, as many writers have noted, developing countries contain a disproportionately large share of the world's biodiversity.[44]

These are very difficult issues that need careful discussion. A full defense of the value of integrity, and hence of the demanding view, would require addressing issues of the sort indicated. I assume, further, that there will very likely be issues of justice relating to each of the domains noted at the beginning of this chapter. This adds more challenge and complexity to the defense of the demanding view.

IV. CONCLUSION

One of the attractions of the demanding view is that it does not regard human well-being as decisive when we are trying to sort out what sustainability might mean. It thus presents perhaps the most radical challenge to the framing of food justice. For we could conceivably achieve sustainability according to the first two views, and food justice in relation to these views, while nonetheless failing to preserve or restore integrity. If the first two views are found wanting in this respect—say, because they are compatible with a much diminished natural world—this would be a reason to take the third view seriously.

NOTES

1. I am grateful to Robert McKim and Mike Doan for conversation about the ideas and issues discussed in this chapter. I also wish to thank Jill Dieterle, who offered helpful suggestions on an earlier draft of this essay.

2. For a nice statement of this sentiment, see Brian Barry, "Sustainability and Intergenerational Justice," in *Fairness and Futurity: Essays on Environmental Sustainability and Social Justice*, ed. Andrew Dobson (Oxford: Oxford University Press, 1999), 101, 93.

3. Here I agree with Barry, "Sustainability," 93.

4. See Barry, "Sustainability," 101.

5. This view is sometimes called "welfare pluralism." See, for example, Michael Jacobs, "Sustainable Development, Capital Substitution and Economic Humility: A Response to Beckerman," *Environmental Values* 4 (1995): 64.

6. Among economists, I would include (e.g.) Robert Solow, Herman Daly, and Salah El Serafy. Among philosophers, prominent examples include John Rawls (and Rawlsians more generally) and David Miller. I think the perspective of Brian Barry could also be categorized as a defense of a minimalist view. However, both Rawls and Barry say things that could support reading them as sympathetic to something along the lines of the human flourishing account that I discuss in the next section. For relevant discussion, see Robert Solow, "Sustainability: An Economist's Perspective," in *Economics of the Environment: Selected Readings* (3rd ed.), ed. Robert Dorfman and Nancy Dorfman (New York: W. W. Norton and Company, 1993); Herman Daly, "On Wilfred Beckerman's Critique of Sustainable Development," *Environmental Values* 4 (1995), and "Sustainable Economic Development: Definitions, Principles, Policies," in *The Essential Agrarian Reader: The Future of Culture, Community, and the Land*, ed. Norman Wirzba (Lexington: University Press of Kentucky, 2003); Salah El Serafy, "In Defence of Weak Sustainability: A Response to Beckerman," *Environmental Values* 5 (1996); John Rawls, *A Theory of Justice* (rev. ed.) (Cambridge, MA: The Belknap Press of Harvard University Press, 1999), *Political Liberalism* (New York: Columbia University Press, 1993), and *Justice As Fairness: A Restatement*, ed. Erin Kelly (Cambridge, MA: The Belknap Press of Harvard University Press, 2001); David Miller, "Social Justice and Environmental Goods," in *Fairness and Futurity*, ed. Dobson; and Barry, "Sustainability." My characterization of the minimalist view is indebted to the discussion in Alan Holland, "Sustainability: Should We Start From Here?" in *Fairness and Futurity*, ed. Dobson; Bryan G. Norton, "Intergenerational Equity and Sustainability," in *Searching for Sustainability: Interdisciplinary Essays in the Philosophy of Conservation Biology* (Cambridge: Cambridge University Press, 2003), 425–32 (in particular), and "What Do We Owe the Future? How Should We Decide?" in *Searching for Sustainability*, 494–500; also Bryan G. Norton and Michael A. Toman, "Sustainability: Ecological and Economic Perspectives," in *Searching for Sustainability*, 227–36.

7. See, for example, Solow, "Sustainability," 181. This statement carries with it two assumptions. First, I attribute to the minimalist view a premise of fundamental equality between all human beings, as I can see no non-question-begging argument for denying such a premise. (Here I agree with Barry, "Sustainability," 96–97.) Second, I assume that location in space and time should not affect a person having legitimate welfare claims. This accords with the views of a number of philosophers who write from otherwise different normative perspectives. See, for example, Barry, "Sustainability"; James P. Sterba, "Global Justice for Humans or For All Living Beings and What Difference It Makes," *The Journal of Ethics* 9 (2005); and Peter Singer, *Practical Ethics* (3rd ed.) (Cambridge: Cambridge University Press, 2011).

8. The maintenance of critical natural capital distinguishes so-called strong sustainability views. For defenses of the latter, see Daly, "On Wilfred Beckerman's Critique" and "Sustainable Economic Development"; and Jacobs, "Sustainable

Development." Since the notion of natural capital strikes me as needlessly vague, I suggest conceptualizing the relevant environmental good here in terms of ecological health (which I discuss more fully at the beginning of section III).

9. This is generally true of the economists who discuss sustainability. See, for example, Solow, "Sustainability," 181–82.

10. For relevant critique of the desire or preference satisfaction view, see Barry, "Sustainability," 101–3; John O'Neill, Alan Holland, and Andrew Light, *Environmental Values* (New York: Routledge, 2008), 21–23, 54–57, 189–95; and Richard Kraut, *What Is Good and Why: The Ethics of Well-Being* (Cambridge, MA: Harvard University Press, 2007), 92–120.

11. See James P. Sterba, *How to Make People Just: A Practical Reconciliation of Alternative Conceptions of Justice* (Totowa: Rowman & Littlefield Publishers, 1988), 45–46. Consider also the list of Rawlsian primary goods, or a Nussbaum-style list of central capabilities, both of which suggest how the relevant needs/capabilities might be specified.

12. Solow, "Sustainability," 181.

13. For further discussion of substitutability, see Alan Holland, "Substitutability: Or, Why Strong Sustainability is Weak and Absurdly Strong Sustainability is Not Absurd," in *Valuing Nature? Ethics, Economics and the Environment*, ed. John Foster (New York: Routledge, 1997), 121–26 (in particular); Norton and Toman, "Sustainability," 227–33; and Norton, "Intergenerational Equity," 425–32.

14. See O'Neill et al., *Environmental Values*, ch. 11 (in particular). This work offers one of the most thoughtful and interesting defenses of a human-flourishing view in the literature. For another notable defense, see Martha Nussbaum, *Women and Human Development: The Capabilities Approach* (Cambridge: Cambridge University Press, 2000) and *Frontiers of Justice: Disability, Nationality, Species Membership* (Cambridge, MA: The Belknap Press of Harvard University Press, 2006).

15. The objective state view is essentially the same as what Parfit calls an "objective list" view. See Derek Parfit, *Reasons and Persons* (Oxford: Oxford University Press, 1984), 493, 499–502.

16. O'Neill et al., *Environmental Values*, 25, 194.

17. This conception of flourishing and harm is indebted to David Wiggins, *Needs, Values, Truth: Essays in the Philosophy of Value* (3rd ed.) (Oxford: Oxford University Press, 1998), Essay I.

18. For further discussion of this type of view, and its relevance for environmental ethics, see J. Michael Scoville, "Historical Environmental Values," *Environmental Ethics* 35 (2013).

19. O'Neill et al., *Environmental Values*, 39, 66, 176, 196–99.

20. Ibid., 196.

21. Ibid.

22. Ibid., 195.

23. For discussion, focused primarily on the case of industrial shrimp production in southern India, see Vandana Shiva, *Stolen Harvest: The Hijacking of the Global Food Supply* (Cambridge: South End Press, 2000), ch. 3.

24. Would this problem be solved if the industrial pattern produced benefits to the economy at large, thereby enabling the local government to redistribute wealth so as

to compensate those, such as the traditional fishing communities, who are harmed by the industrial pattern? This would be better than nothing. But even in this case, there would seem to be particular harms that simply would not, and could not, be compensated for. O'Neill et al.'s flourishing view helps illuminate the relevant harms here.

25. Assuming compensation is actually paid; it often isn't in cases of this sort.

26. A complexity here is that in some cases life might become easier for the people in question, and they might welcome this.

27. For discussion of both types of negative effect, see Daly, "Sustainable Economic Development" and Vandana Shiva, "Globalization and the War against Farmers and the Land," in *The Essential Agrarian Reader*, ed. Wirzba, 121–39.

28. See Vandana Shiva, "The Impoverishment of the Environment: Women and Children Last," in *Environmental Philosophy: From Animal Rights to Radical Ecology* (4th ed.), ed. Michael Zimmerman et al. (Upper Saddle River: Pearson Education, Inc., 2005), 180.

29. For relevant discussion of subsistence and agrarian communities in India and Latin America, see Ramachandra Guha and Juan Martinez-Alier, *Varieties of Environmentalism: Essays North and South* (London: Earthscan Publications Ltd., 1997).

30. My discussion here is indebted to Kyle P. Whyte, "Food Justice and Collective Food Relations," forthcoming in *The Ethics of Food: An Introductory Textbook*, ed. Anne Barnhill et al. (Oxford: Oxford University Press, 2015).

31. On the good of health, see O'Neill et al., *Environmental Values*, 192–94.

32. The relationship between these two goods is complex: for example, many people might (and often do) forego a fully nourishing diet (assuming they are not, on that count, seriously *malnourished*) in order to maintain social affiliation in ways they deem valuable.

33. Whyte, "Food Justice," 5.

34. The sense I won't discuss refers to those environments that are, as O'Neill et al. put it, necessary for "maintaining the capacity to appreciate the natural world and to care for other species" (*Environmental Values*, 195). This capacity is understood to be an objective good partly constitutive of human flourishing.

35. Ibid.

36. Whyte, "Food Justice," 8–9.

37. The authors seem to recognize, but don't respond to, the problem here. They write: "Because autonomy, the capacity to govern one's own life and make one's own choices, is a human good, it may matter that those objective goods be endorsed by a person. One cannot improve an individual's life by supplying resources that are valuable to the individual by some objective criterion, but not in light of the conception of the good life recognised and accepted by that individual: a person's life cannot go better in virtue of features that are not endorsed by the individual as valuable" (O'Neill et al., *Environmental Values*, 25).

38. Leopold defined what he called "land health" as "the capacity of the land for self-renewal" (see Aldo Leopold, *A Sand County Almanac and Sketches Here and There* [New York: Oxford University Press, Inc., 1949], 221). This characterization maps onto the first property I note above. Leopold says other things that suggest the second property as well (see, e.g., the discussion in the section entitled "The Land Pyramid," 214–20). For helpful discussion concerning the conceptualization

of ecological health, see J. Baird Callicott, "The Value of Ecosystem Health," in *Beyond the Land Ethic: More Essays in Environmental Philosophy* (Albany: State University of New York Press, 1999); and David J. Rapport, "Ecosystem Health: More than a Metaphor?" *Environmental Values* 4 (1995), and "Sustainability Science: An Ecohealth Perspective," *Sustainability Science* 2 (2007). I am indebted to Callicott and Rapport, in particular, in my characterization of ecological health in the text above.

39. See James R. Karr, "Ecological Integrity and Ecological Health Are Not the Same," in *Engineering Within Ecological Constraints*, ed. Peter C. Schulze (Washington, D.C.: National Academy Press, 1996), and "Health, Integrity, and Biological Assessment: The Importance of Measuring Whole Things," in *Ecological Integrity: Integrating Environment, Conservation, and Health*, ed. David Pimental et al. (Washington, D.C.: Island Press, 2000); also Paul L. Angermeier and James R. Karr, "Biological Integrity versus Biological Diversity as Policy Directives," *BioScience* 44 (1994). A variety of conservation thinkers and philosophers have endorsed Karr's view, or something like it.

40. See Karr, "Ecological Integrity," 212; also Alan Holland, "Ecological Integrity and the Darwinian Paradigm," in *Ecological Integrity*, ed. Pimental et al., 51.

41. Regarding "wilderness" (as the base-datum for Leopold's conception of land health), the ecologist David Rapport remarks: "There may be no reason to accept in all cases that *a priori* wilderness is healthy in the broad sense of being supportive of human health and economic activity" (Rapport, "Ecosystem Health," 297).

42. For discussion of nature's otherness, see Bernard Williams, "Must a Concern for the Environment be Centred on Human Beings?" in *Making Sense of Humanity and Other Philosophical Papers* (Cambridge: Cambridge University Press, 1995), 237–40; and Robert Elliot, *Faking Nature: The Ethics of Environmental Restoration* (London: Routledge, 1997), 59–62.

43. There are complexities here relating to the question of scale. For example, health ought not to be compromised at a large or nontrivial spatial scale, while it might be justifiably compromised at a more local spatial scale. This issue obviously requires more discussion.

44. See, e.g., Michael Wells, "Biodiversity Conservation, Affluence and Poverty: Mismatched Costs and Benefits and Efforts to Remedy Them," *Ambio* 21 (1992): 237; and Mark Dowie, *Conservation Refugees: The Hundred-Year Conflict between Global Conservation and Native Peoples* (Cambridge: MIT Press, 2009), xxvii.

Chapter 2

Food, Hunger and Property

Stephen Minister

With the end of the timeframe for the Millennium Development Goals, we are hearing stories of the significant strides that have been made over the last two decades in the fight against poverty and hunger.[1] While some countries have made real progress that is worthy of celebration and emulation, this should not obscure the fact that in recent decades the general trend has been only a modest decline in the number of people who do not have enough food to eat.[2] Moreover, this trend has been fragile and prone to reversal. Hence, even with the recent progress, there is still much work to be done on the issue of hunger, and not just in the developing world. Indeed, in the United States, 19 percent of households could not afford to buy food they needed at some point in 2013.[3] More than one of every five American children lives in one of these food insecure households.[4] Globally, one in four children is stunted, a key symptom of inadequate nutrition, and overall one out of every nine people (about 805 million human beings) regularly go to bed hungry.[5] Throughout the world, almost half of all child deaths are attributable to undernutrition,[6] and in total around 25,000 people die each day from hunger, malnutrition and related diseases.[7] That amounts to more than eight times as many people each day as died in the September 11 terrorist attacks and more people than died each day in the Holocaust and the Rwandan genocide combined. It goes without saying that the outrage generated by those last three events has no parallel in our response to hunger. With so many children and adults wrestling day after day with unsatisfied hunger pangs, and so many succumbing to the effects of inadequate access to food, where is the urgency to address this problem?

Perhaps one reason for the lack of urgency is the persistent tendency to think of hunger in a Malthusian way. There are simply too many people and not enough food, so hunger is a biological inevitability. On this view, hunger is unfortunate and we should do what we can to decrease it, but it isn't really anyone's fault the way those other catastrophes were. At certain times in the past, this view may have been correct, but in the modern world it is the wrong way to think about hunger.

One clear indicator that there is more at work than just biology is the fact that hunger and malnutrition are not randomly distributed, but are concentrated among certain groups. Globally, there are 160 million more women who are undernourished than men.[8] In Guatemala, almost 80 percent of children in indigenous families are stunted, whereas only 40 percent of children in nonindigenous families are.[9] These patterns are not the results of biological or environmental bad luck, but social practices and political policies that have made and continue to make certain groups economically vulnerable and food insecure. Contemporary hunger is a sociopolitical phenomenon, not a biological one. In the modern world, hunger is not just about the relation between a person and food, but about the relations between people.

Amartya Sen has made this point regarding famines, noting that sometimes countries continue to export food while certain people within the country are starving.[10] Such cases make clear that hunger isn't due to a lack of food, but a lack in the ability to procure food, through economic exchange or otherwise. Why though do certain groups systematically lack the ability to procure enough food? Economic theory alone cannot explain this. Instead we must look to social and political power relations, including the power to make economic and social policy. Hunger is not the result of allegedly anonymous market forces anymore than it is simple biology. It is instead the result of specific policies and practices, adopted and perpetuated because they benefit certain powerful interests. Thus as Jenny Edkins has pointed out, "starve" is usually a transitive verb. It is something one group of people does to another.[11] Jean Ziegler, former UN Special Rapporteur on the Right to Food, is even more direct. He writes: "The destruction, every year, of tens of millions of men, women and children from hunger is the greatest scandal of our era. . . . In its current state, the global agricultural system would in fact, without any difficulty, be capable of feeding 12 billion people. . . . Hunger is thus in no way inevitable. Every child who starves to death is murdered."[12]

When trying to deny one's involvement in the deaths of others, it is always best to have a good alibi. At present, one important source of alibis for denying complicity in the starvation or malnutrition of others is economic rationality.[13] Because of the prestige currently afforded to economic rationality, it has become common for appeals to market logic or narrowly economic goods to justify government policies (or government inaction) which foreseeably and avoidably deprive some people of access to adequate nutrition. So, for example, in the United States the Institute of Medicine has judged existing levels of government food assistance to be inadequate.[14] And yet, despite the enormous wealth within the United States, even the existing levels are perpetually at risk of being cut in the name of fiscal responsibility.

Internationally the priority of economic rationality over basic human needs shows up in a host of ways. The bananas for sale in my local grocery story are usually from Guatemala, a country where half the children are malnourished. Why do those bananas end up in the stomachs of middle-class people in my neighborhood rather than hungry children in the communities in which they are grown? The short economic answer is that there is greater "demand" for the bananas in the United States. Economic "demand" does not refer to who wants the bananas more, nor who needs them more, but simply who has the ability and willingness to pay more for them. It is only by translating demand into dollar signs that my neighbors can be said to have greater "demand" for the bananas than malnourished children. This dominance of economic demand over human need is demonstrated by the fact that a strong majority of low-income countries are net exporters of agricultural goods.[15] That is, despite most of these countries having high levels of malnutrition, they ship more crops away from their malnourished children than they bring back into their country.

Certain global institutions, like the World Trade Organization (WTO) and the International Monetary Fund (IMF), are strongly committed to economic rationality and so have sometimes pushed poor countries to adopt policies, justified by economic rationality, that have increased hunger and malnutrition. For example, in the 1980s and 1990s the IMF and the World Bank pressured most poor countries to adopt structural adjustment programs that required significant cuts in government spending. In retrospect the World Bank concluded that these cuts, including cuts to agricultural and social programs, "increased food insecurity among many rural households."[16] In the early 2000s, the IMF pushed Malawi and Niger, countries plagued by recurrent famine, to sell off their emergency food stockpiles because they were deemed to distort the market. Foreseeably and avoidably, when crops next failed the governments were inadequately prepared, leading to unnecessary suffering and death.[17] When children are famished so that markets can be free, something has gone terribly wrong.

The IMF and WTO have also pushed trade liberalization on the basis of projected aggregate economic gains. Yet in poor countries, all too often the economic gains go to large-scale commercial farms and agribusinesses, while the costs are imposed on smallholder farmers. Since many malnourished children are in smallholder farm families, the costs are often borne by those who can least afford them. While economic theory and metrics cannot simply be ignored, neither can they justify ignoring the cries of a child who must go to bed hungry.

It is important to recognize that in none of these cases did economic rationality, by its own power, force these decisions. Instead, the decisions were taken based on political power, with economic rationality functioning as the legitimating reason for the decision. In each of these cases, certain parties benefited from the decision. For example, banks benefited from the austerity programs that prioritized repayment of loans—some of which were surely odious—over people having access to adequate nutrition. All governments defy economic rationality in certain ways. Thus, we should not see these examples as simply the inevitable outcomes of economic logic, but instead as the outcomes of political decisions, made under the cover of economic rationality.[18]

Economic rationality is able to function as an alibi for perpetuating hunger because it is focused on economic goods (e.g., growth, productivity, efficiency) that are detached from a broader account of justice and the human good. Thus, what we need is a recontextualization of economic goods and rationality within an account of justice and the human good. This task goes well beyond the space I have here, so this chapter focuses on recontextualizing a concept that is foundational for economics, namely, property. I pursue this task by analyzing three thinkers who, though quite different, share an overlapping consensus on the view that the need for basic goods, especially food, can trump property claims. The three thinkers are Thomas Aquinas, John Locke and Emmanuel Levinas. These thinkers are important representatives of intellectual traditions—Christian theology, modern liberalism and postmodernity—that continue to have significant influence. Though writing from very different philosophical perspectives and historical contexts, these three thinkers all agree that the existence of property is justified precisely because it serves the satisfaction of human needs. However, as such, property claims lose their justification in those cases where they result in depriving humans of the things they need. Hence, human needs for the means of subsistence, including especially food, trump property rights.

What these thinkers point us to is not the rejection of the notion of property or economic rationality, but instead putting them at the service of human needs. As such, they all take up a middle position between those (like Proudhon and Marx) who claim that private property is the problem and those (like Nozick and Rand) who tend to absolutize property rights. It is my hope that in a similar way, these thinkers will point us toward a view of economic rationality as neither simply the problem, nor simply the solution, but as a valuable tool that finds its highest calling in serving justice and the human good. People were not made for economic rationality, but economic rationality for people.

I. AQUINAS

Aquinas was a thirteenth-century Italian thinker who has had a profound influence on Catholic theology. His writings on property occur in the charged context of a debate about the possibility and desirability of apostolic poverty, that is, of renouncing all worldly possessions. The thirteenth century saw the rise of mendicant orders, such as the Franciscans seeking to emulate the example of St. Francis of Assisi. Some Franciscans, like the later William of Ockham, offered intellectual support for their renunciation of property by arguing that property ownership was not natural, but was instead a result of humanity's fall into sinfulness.[19] On this account, it became necessary after the fall for humans to create laws assigning and protecting property claims in order to mitigate the pernicious effects of human greed. Property ownership thus exists only because of human sinfulness.

This view caused a variety of tensions throughout the Catholic Church, itself a rather large property owner, and was ultimately ruled heretical by Pope John XXII in the fourteenth century. John ruled that property was not the result of human law, but was part of the natural order, such that even Adam and Eve had legitimate property claims over the things they used.[20] Aquinas, writing before this pronouncement, had attempted to craft a middle ground between these two views.

Aquinas starts from the view that "God has sovereign dominion over all things," that is, that ultimately everything belongs to God. For Aquinas, God is not selfish, but has "directed certain things to the sustenance of humans' bodies. For this reason, humans have a natural dominion over things as regards the power to make use of them."[21] That is to say, it is natural—which here means in accordance with the way God has created the world to be—for humans to eat food as nourishment for our bodies. Hence, it is natural and good for us to *use* appropriate items for this purpose. So far, neither Ockham nor Pope John XXII would find much to disagree with. But here's the crux of the issue: does the fact that it is natural and good to appropriate food to oneself establish a natural property claim? Aquinas holds the view, which Ockham will subsequently defend, that it does not.

Instead, Aquinas claims that "there is division of possessions, not according to the natural law, but rather according to human agreement, which belongs to positive law. . . . Hence, the ownership of possessions is not contrary to natural law but an addition thereto devised by human reason."[22] Aquinas is here anticipating Ockham's claim that property is the result of human law, not nature, but Aquinas has a very different reason for thinking this. Property ownership is not an expression of fallen humanity, nor even a protection against fallen humanity, but is instead a rational addition to natural law. Aquinas thinks property ownership is rational because, by assigning objects to the possession of individuals, ownership promotes taking care of the objects and is more conducive to order and harmony.[23]

On the one hand, the fact that property ownership is rational means that it is good and right. On the other hand, the fact that it is not natural means that property always exists within a prior context in which it is subordinate to natural law. Since it is natural for humans to use resources for sustenance, claims of property ownership must be subordinated to people's need for food. As Aquinas puts it: "The division and appropriation of things which are based on human law do not preclude the fact that [human] needs have to be remedied by means of these very things. Hence, whatever goods some have in superabundance are due, by natural law, to the sustenance of the poor."[24] This is a very strong claim, analogous to those that Peter Singer makes regarding poverty.[25] Because of the priority of natural human needs over rational property ownership, property claims must yield to human needs. Hence, if we have things in superabundance, that is, beyond our own needs, we are obligated by natural law to give them to those in need.

For Aquinas, the primary expression of this is an ethical obligation to charity. Those with wealth should, through charitable giving, provide for the needs of the poor. But what happens if the wealthy are unwilling to do so? Would it be morally acceptable for a person who is poor to take food from a person with abundance? Aquinas is seemingly between a rock and a hard place since he is committed to the priority of human need over property claims, but is also committed to the commandment "Thou shall not steal." Ever clever at navigating tight spots, Aquinas reasons that since human need trumps property claims, property claims dissolve in the face of urgent need and with them the very possibility of stealing. "If the need be so manifest and urgent . . . then it is lawful for a [person] to succor his own need by means of another's property, by taking it either openly or secretly, nor is this properly speaking theft."[26] When one is in real need, it is morally acceptable to take food from a person with excess, not because it is acceptable to steal when in need, but because when in need, it is not even stealing. It is instead using food the way God intended and so is in accordance with the natural law. Aquinas also explicitly permits Robin Hood cases, in which an intermediary takes from the wealthy to give to the poor.[27]

Finally, Aquinas endorses legal provisions to ensure that those in need have access to food and other necessary resources. It is acceptable and necessary that public authorities appropriate resources from their subjects in order to provide for the common good.[28] In his analysis of Old Testament law, Aquinas praises the legal provisions designed to ensure that all have access to food, such as rules requiring farmers to allow the poor to collect food from their fields at certain times.[29] Hence, for Aquinas, ensuring that everyone has access to adequate nutrition is a matter of both justice and charity. Property claims, though rational and good, must always be seen in the broader context of the human good and as such must yield in the face of human needs.

II. LOCKE

Locke is writing amid shifting political power and burgeoning capitalism, both of which he was active in and both of which he was eager to defend philosophically. However, despite the reputation of Locke's *Second Treatise* as offering a rights-based theory of government, he begins his discussion of the state of nature not with a claim to rights, but with the ascription of duties. No doubt as a counterpoint to Hobbes, Locke gives an extended argument for the natural moral equality of persons, on account of which we find ourselves obligated to others from the beginning. This natural moral equality leads Locke to posit a strong duty to preserve the lives of others. He writes: "Everyone, as he is *bound to preserve himself*, . . . so by like reason, when his own preservation comes not in competition, ought he, as much as he can, *to preserve the rest of mankind.*"[30] Taken literally, this is a very demanding claim—that we ought to do as much as we can, short of endangering our own self-preservation, to preserve the lives of others. This is again reminiscent of Peter Singer's work. Locke is arguing not just for a negative duty to do no harm, but for a positive duty to actively help others in need *whenever* we can.

When Locke does introduce rights, he explicitly states that the right to self-preservation entails the right to the means of subsistence. According to Locke, "natural reason" tells us that humans, "being once born, have a right to their preservation, and consequently to meat and drink, and such other things as nature affords for their subsistence."[31] While those of us who are vegetarians prefer a right to tofu and drink, it is clear that Locke holds there is a natural right to food and other necessary means of subsistence. It is in fact this natural right that provides the moral justification of property rights. Locke reasons that in eating an apple, I am removing that apple from the common endowment of humanity since it is no longer available for others, and thus I am claiming it for myself. This is the first and most fundamental property claim. Since meeting one's subsistence needs, especially the need for food, requires appropriating goods from the commons to oneself, the right to the means of subsistence founds the right to property.

It is essential to recognize that for Locke, what justifies the *existence* of property rights is not the mixing of labor, but the obligation to preserve human life. It is only after Locke establishes that there are property rights that he inquires into *how* it is possible to appropriate a particular object to oneself from the commons. It is this question to which the account of mixing one's labor furnishes the answer. The capacity to mix one's labor with natural resources does not justify the institution of private property, but simply explains how one comes to own particular objects. Labor is the *how*, not the *why*, of property. The right of each and all to the means of subsistence is the grounding on which the notion of property rights is justified. We have to eat: that is why there are property rights.

This foundational right to the means of subsistence provides not only justification for the notion of property rights, but also limits thereto. These bounds, or limits, are the well-known Lockean provisos that we can only appropriate what we can use before it spoils and that we must leave as much and as good for others. These provisos indicate that our property claims are delimited by the obligations we have to others. Famously, C. B. Macpherson and others have argued that these limits are undercut by the introduction of money.[32] Locke, of course, claims that the institution of money is a natural development to which all tacitly consent. Since money does not spoil it makes possible unlimited accumulation and thus gross inequality. Since it is food and not money that is directly necessary for the preservation of life, the subsistence needs of others do not seem to delimit monetary accumulation and inequality.

While there is no doubt that Locke is eager to portray accumulation and inequality as natural, justified and (quite implausibly) consented to by all, this argument does not undercut our obligation to provide aid for those in need. Locke repeatedly reminds us that money is not like food or other basic goods. Some critics read this claim as part of the argument that the right to accumulate wealth is independent from and so can override the right to food. However, these reminders could also be read as Locke's recognition of the fact that the right to accumulate wealth can never override the right to food. There is a natural necessity to food that money can never displace because food is a fundamentally different type of good. It is not just one commodity among others. This latter reading is more consistent with the other claims Locke makes regarding food and money throughout his writings.

First of all, there is nothing in Locke's account of money that diminishes the right of all to the means of subsistence and the duty of all to assist those in need when possible. In fact, one of Locke's famous (or infamous) arguments to justify the existence of money and the possibility of accumulation is that it leads to increased productivity which, in turn, allows for the needs of many to be met. While this argument may be overly simplistic, it demonstrates that the subsistence needs of the many continue to be a justifying concern even after the introduction of money.[33]

Moreover, the idea that property rights hold priority over subsistence rights for Locke runs into insuperable difficulties when we consult works beyond the *Second Treatise*. In these works, Locke clearly prioritizes subsistence needs over property rights in private, commercial and public contexts. In the

First Treatise, Locke even uses the language of entitlement for those in extreme need. He writes:

> God . . . has given no one of his Children such a Property, in his peculiar Portion of the things of this World, but that he has given his needy Brother a *Right* to the Surplusage of his Goods; so that it cannot justly be denied him, when his pressing Wants call for it . . . *Charity* gives every Man a Title to so much out of another's Plenty, as will keep him from extreme want, where he has no means to subsist otherwise.[34]

This passage is a clear echo of Aquinas's claim that the right of those in need to the means of subsistence engenders an obligation to private charity for those with a superabundance of goods.

In his essay "Venditio," Locke again demonstrates the priority of subsistence rights, this time in a commercial context. In "Venditio," Locke reflects on the question of fair pricing of goods for sale. Contrary to Aquinas, he argues that, in general, the market price—whatever it is—is just, so long as there is no fault in the product sold and the price is the same for all customers. While Locke criticizes taking advantage of an individual's ignorance or desperation in order to charge that individual more, he nonetheless allows that in the midst of a general food shortage it is permissible to charge more for food since the market price will naturally go up. However, if the market price exceeds what those in extreme need can afford, then the right to food trumps the seller's right to property. Locke writes:

> But though he that sells his corn in a town pressed with famine at the utmost rate he can get for it does no injustice against the common rule of traffic, yet if he carry it away unless they will give him more than they are able, or extorts so much from their present necessity as not to leave them the means of subsistence afterwards, he offends against the common rule of charity as a [hu]man, and if they perish any of them by reason of his extortion [he] is no doubt guilty of murder.[35]

Very clearly, the rights to property and to free market transactions are here trumped by the right of those in need to food. Locke's language here is so strong that he describes the failure to help as not just letting die, but actively killing, that is, murder. One wonders what he would make of the situation of low-income countries with malnourished children that are net exporters of agricultural goods.

It is not clear from this passage whether Locke regards such an action as legally equivalent to murder or only morally equivalent. Nonetheless, in his essay proposing reforms to England's Poor Law, he demonstrates his belief that the moral obligation to aid those in need should be translated into legal obligations.[36] He supports a welfare program in which those with resources are assessed a tax which is used to aid those in need. He also recommends a school lunch program and an economically based affirmative action scheme in which those with large industrial or agricultural enterprises are required to offer apprenticeships to youths from poor families. Finally, he suggests that parishes be legally punished if any of their people die due to lack of necessary aid. These recommendations point to Locke's belief that the state has a role to play in creating a social and economic system in which all are able to meet their basic subsistence needs. While it is clear that Locke believes a strong private property regime should be part of that system, the right of all to the means of life is fundamental.

As with Aquinas, Locke holds that there are responsibilities of both charity and justice to ensure that people have access to food. Like Aquinas, this is due to the priority of human need over property claims. Unlike Aquinas, Locke has enthusiasm for the burgeoning capitalist system, including market pricing, profit making, interest rates, investment opportunities, accumulation and inequality. And yet this enthusiasm does not undermine Locke's fundamental commitment to ensuring that all can meet their basic needs. Instead, he suggests that this system and every system have an irrevocable responsibility to attend to the basic needs of all. For when describing the extent of legislative power in the *Second Treatise* he writes: "The Law of Nature stands as an eternal rule to all, . . . *legislators* as well as others. The *rules* that they make for other [people]'s actions must . . . be conformable to the law of nature . . . and *the fundamental law of nature being the preservation of mankind*, no human sanction can be good, or valid against it."[37]

III. LEVINAS

We now turn to Levinas, a twentieth-century Lithuanian-born, French Jewish phenomenologist. As a phenomenologist, Levinas rejects the abstract, disembodied conception of the subject prevalent in much modern philosophy. One of the concrete physical experiences Levinas regularly mentions to emphasize human embodiment and particularity is the need to eat. One of Levinas's favorite ways to make this point is with the phrase: "A famished stomach has no ears."[38] For Levinas this phrase is not just an empirical point about the difficulty of listening attentively while hungry, but is also the deeper point that the potentials of human consciousness depend on meeting basic physical needs. Mind over matter lasts only as long as one's stomach allows. Our need for food is never dissolved or overcome through rational mastery. Hunger remains always a reminder of our material dependence, a permanent vulnerability. As such, embodiment, with all its attendant needs, is not a contingent or trivial fact for humans, but is essential to our existence. Food should never be an afterthought when contemplating human life, since it is in a very real way the core of our being. As Levinas puts it, "Eating does not reduce itself to the set of gustative, olfactory, kinesthetic, and other sensations that would constitute the consciousness of eating . . . in satiety the real I sink my teeth into is assimilated, the forces that were in the other become my forces, become me."[39] Levinas is here rejecting the reduction of food to conscious experience of food as well as the instrumental representation of food, that is, food as mere fuel for the human machine. I do not merely use the food I consume; in a very real way, it becomes me. As such, food is not like other objects or pieces of property whose value can simply be determined by the markets any more than the value of a human being, a human body, can be.

Whereas Locke relates the basic human need for food to the economic and political changes in his day, Levinas relates them to the evolving scientific approach to the world, including the practices of the social sciences. For Levinas, scientific rationality has become a dominant force reshaping human society in the twentieth century and beyond. However, he does not think the pursuit of scientific understanding is simply an unbiased, neutral investigation grounded in the desire for truth for truth's sake. It is instead shaped by the lived human condition and grounded in human need. Our material dependence leads to a desire to control the material world, to bend its powers to the fulfillment of our needs. It is this desire for control that motivates the desire to understand the physical world and thus underlies modern science. We might think here of Francis Bacon's famous claim that "knowledge is power." Levinas's favored reference is to a character in the French renaissance writer Francois Rabelais's work named "Messer Gaster," literally "Mr. Stomach." In Rabelais's work Messer Gaster is referred to as the "first Master of Arts in the World."[40] For Levinas, the pursuit of knowledge and control is a key part of the modern project of freedom. Since physical needs are an essential part of our being, freedom cannot be the absence of need, nor a rational agency independent of physicality, but must involve the mastery of our material dependence, the mastery of those things necessary to satisfy our needs.

On the one hand, Levinas supports the pursuit of scientific understanding and innovation on account of its capacity to satisfy physical needs and enhance our freedom. On the other hand, he is wary of the limits of the mindset that scientific investigation encourages. Scientific investigation requires an objectification of the world, an attitude that approaches the world as a collection of objects to be studied and manipulated. While this mindset is quite useful for most material things, Levinas is wary of the dangers of the objectification of human beings. While humans are essentially material bodies, objective material functioning does not exhaust the significance of our humanity. As a Jew whose family fled persecution in Lithuania and whose parents and brothers were killed in the Holocaust, Levinas is all too aware that the desire for freedom and control supplemented by the objectification of others can become a justification for suffering and death.

Because of this background, Levinas is very interested in the experiences we have of others that defy objectification. One type of experience that Levinas focuses on is the experience of others in need, including in need of food. "Hunger," Levinas writes, "is strangely sensitive in our secularized and technological world to the hunger of the other [person]. All our values are worn out except this one. The hunger of the other awakens [people] from their sated drowsing."[41] In the experience of another person in need of food, we find our project of freedom and control called into question and subordinated to the responsibility to *do* something. As our own hunger calls attention to the significance of our embodiment, the hunger of others calls our attention to the significance of their embodiment and thereby calls for a tangible, material response, for actions that go beyond our social and political theorizing and that may even call them into question. As Levinas eloquently puts it,

> [Hunger] is a privation whose acuity consists . . . in not being able to console oneself with a "spiritually ordered" world in which historians find the wretched sated by the odor of roasts, paid with the sound of coins, and consoled by their consciousness of the harmony of the social body as a whole, and by the perfect definition of their status. The hunger which no music pacifies secularizes all this romantic eternity.[42]

Hunger cannot be stilled by theories about the greater good, a rationally ordered world, the naturalness of property rights, or the virtues of free markets and free trade. "Hunger," Levinas tells us, "is deaf to every reassuring ideology."[43] The stomach without ears can only be satisfied with the concrete provision of food.

In one of the bluntest statements in all of *Totality and Infinity*, Levinas asserts that "to leave [people] without food is a fault that no circumstance attenuates."[44] This is a bold and unyielding statement of responsibility, yet quite consistent with the obligation to those in need propounded by Aquinas and Locke. To leave others without food is to deprive them not simply of property, but also of their very bodies, their very selves, their very human potential. Because of this, for Levinas, as for Locke and Aquinas, our property claims must yield in the face of hungry others. "The problem of a hungry world can be resolved only if the food of the owners and those who are provided for ceases to appear to them as their inalienable property. . . . Scarcity is a social and moral problem and not exclusively an economic one."[45] In this prioritization of hunger, property rights are not abolished, but are simply subordinated to the material needs of others. This point is echoed by Levinas's oft-repeated mantra that we must be willing to take the bread out of our own mouth to feed hungry others.[46] In so doing, we are not only giving of our property, but also of our very selves. Levinas calls on us to be willing to "take the initiative of renouncing [our] rights so that the hungry can eat."[47]

IV. CONCLUSION

Despite their disparate contexts and intellectual commitments, Aquinas, Locke and Levinas exhibit an overlapping consensus on the fundamental importance of ensuring that everyone has access to the means of subsistence, including especially sufficient food to eat. This gives rise to a responsibility that trumps property claims and, by extension, the economic rationality built upon them. That said, such a responsibility does not always translate neatly or clearly into practical actions or policies. In the concrete, figuring out how to ensure sustainable access to adequate nutrition for all involves grappling with particular, complex situations. As such, what Aquinas, Locke and Levinas offer us today are not universal policy prescriptions, but a prophetic call. They remind us of what goodness and justice call us to in the midst of imperfect political and economic systems that cause real human suffering and death. They remind us of what holds fundamental importance in the midst of our favored ideologies that, in attempting to justify these systems, function as justifications of the suffering and death.

That said, there are a couple practical points we can draw from their perspectives. By conceiving of hunger as primarily a sociopolitical problem, they provide a challenge to the tendency to think that the solution to hunger is simply increased agricultural productivity. While increasing productivity may make certain issues easier to resolve, by itself increased productivity will not solve the underlying social injustices. This is especially true when increasing productivity means shifting to more capital-intensive forms of agricultural, which can exacerbate the situation of smallholder farmers who are already prone to food insecurity. From a food security perspective, it may be preferable to invest in smallholder agriculture to improve the quality of inputs, farming techniques, storage capacity and market access, as well as investing in education and job creation to make possible alternatives to smallholder farming. A number of developing countries have also implemented cash transfer programs that have helped decrease food insecurity.[48] While countries can learn much from each other, the best mix of policies to address food insecurity will depend on the particular social, political, economic and environmental circumstances of each country.

At the international level, we need to reorient multilateral institutions and bilateral relationships to focus on ensuring that everyone has access to the means of satisfying their basic needs. While liberalizing markets and trade may at times be a part of that, liberalization should not be pursued as an end in itself, but only when and to the degree that it serves human well-being, especially the well-being of the most vulnerable. This is not to deny the value of economic rationality, nor the reality of economic forces, resource scarcity and the necessity of economic sustainability. But since economic forces and the rationality based on them are to a large degree created by social practices and political policies, we can and should modify them to respond to unmet needs. Ultimately, this requires increasing the political power, at both national and international levels, of marginalized groups that are prone to food insecurity. Only as empowered co-creators of economic and social policy can they ensure that their voices will not be ignored in favor of narrow economic rationality and the property claims of the wealthy.

As long as children in our neighborhoods and around the world go to bed hungry, we have more work to do. Aquinas, Locke, and Levinas may not give us all the economic or political answers to these problems, but what they do offer us is the powerful reminder that these are problems, *fundamental* problems, and thus working toward their solutions should be of the utmost importance.

NOTES

1. I would like to thank the Augustana Research and Artist Fund for supporting my work on this essay.

2. For an analysis showing that framing the poverty-related targets in terms of halving proportions, rather than absolute numbers, and backdating the baseline to 1990, significantly lowered the bar for achieving MDG 1, see Thomas Pogge, *World Poverty and Human Rights* (Cambridge: Polity Press, 2008), 11–13.

3. Rebecca Riffkin, "Mississippians' Struggles to Afford Food Continued in 2013," *Gallup*, March 21, 2014, accessed March 11, 2015, http://www.gallup.com/poll/167774/mississippians-struggles-afford-food-continue-2013.aspx.

4. Alisha Coleman-Jensen, Christian Gregory, and Anita Singh, *Household Food Security in the United States in 2013: Statistical Supplement*, U.S. Department of Agriculture, Economic Research Service, AP-066 (2014): 6, accessed March 11, 2015, http://www.ers.usda.gov/media/1565530/ap066.pdf.

5. Childhood stunting data from UNICEF, "Global Malnutrition Trends (1990–2013)," accessed March 11, 2015, http://data.unicef.org/resources/2013/webapps/nutrition. Total chronic undernutrition data from FAO, IFAD, and WFP, *The State of Food Insecurity in the World 2014: Strengthening the Enabling Environment for Food Security and Nutrition* (Rome: FAO, 2014), 4.

6. UNICEF, "Undernutrition Contributes to Half of all Deaths in Children Under 5 and is Widespread in Asia and Africa," last updated February 2015, http://data.unicef.org/nutrition/malnutrition.

7. Roger Thurow and Scott Kilman, *Enough: Why the World's Poorest Starve in an Age of Plenty* (New York: Public Affairs, 2009), xiv.

8. World Food Programme, *WFP Gender Policy* (Rome: WFP, 2009), 3, accessed March 10, 2015, http://documents.wfp.org/stellent/groups/public/documents/communications/wfp203758.pdf.

9. World Bank, "Nutrition at a Glance: Guatemala," accessed on March 9, 2015, http://siteresources.worldbank.org/NUTRITION/Resources/281846-1271963823772/Guatemala.pdf.

10. Amartya Sen, *Poverty and Famines* (Oxford: Oxford University Press, 1981), 40.

11. Jenny Edkins, *Whose Hunger? Concepts of Famine, Practices of Aid* (Minneapolis: University of Minnesota Press, 2000), 58.

12. Jean Ziegler, *Betting on Famine: Why the World Still Goes Hungry* (New York: New Press, 2013), xiii.

13. There are other alibis as well. For example, in the United States, the rhetoric of energy independence is used to justify an inefficient process whereby almost one-third of the U.S. corn crop ends up in gas tanks rather than mouths.

14. Institute of Medicine, "Supplemental Nutrition Assistance Program: Examining the Evidence to Define Benefit Adequacy," January 17, 2013, accessed March 10, 2015, http://www.iom.edu/snapadequacy.

15. Francis Ng and M. Ataman Aksoy, *Who are the Net Food Importing Countries?* (Washington, D.C.: World Bank, 2008), 9, accessed March 11, 2015, http://elibrary.worldbank.org/doi/pdf/10.1596/1813-9450-4457.

16. Michael Morris et al., *Fertilizer Use in African Agriculture: Lessons Learned and Good Practice Guidelines* (Washington, D.C.: World Bank, 2007), 4, accessed March 10, 2015, http://documents.worldbank.org/curated/en/2007/01/7462470/fertilizer-use-african-agriculture-lessons-learned-good-practice-guidelines.

17. John R. Butterly and Jack Shepherd, *Hunger: The Biology and Politics of Starvation* (Hanover, NH: Dartmouth College Press, 2010), 133–38.

18. This point about how economic rationality functions is distinct, at least theoretically, from questions about the motives of those who appeal to economic rationality. I take it there are a range of motivations for such appeals, from cynical to sincere. While some make appeals to economic rationality simply to justify self-serving material interests, others do so because they believe that following economic rationality will lead to better outcomes overall and so is the right thing to do. Surely sincerity is better than cynicism, but just as surely sincerity does not justify unnecessary deprivation.

19. William of Ockham, *A Short Discourse on Tyrannical Government*, trans. John Kilcullen (Cambridge: Cambridge University Press, 1992), bk. 3, ch. 7.

20. For a discussion of the argument of Pope John XXII, see John Kilcullen, "The Origin of Property: Ockham, Grotius, Pufendorf, and Some Others," in *A Translation of William of Ockham's Work of Ninety Days*, trans. J. Kilcullen and J. Scott (Lewiston, NY: Edwin Mellon Press, 2001).

21. Thomas Aquinas, *On Law, Morality, and Politics*, ed. William P. Baumgarth and Richard J. Regan (Indianapolis: Hackett, 1988), *ST* II-II 66.1.

22. Ibid., II-II 66.2.

23. Ibid.

24. Ibid., II-II 66.7.

25. Peter Singer argues for a strong obligation to help those in need on the basis of the claim, "If it is in your power to prevent something bad from happening, without sacrificing anything nearly as important, it is wrong not to do so" (Peter Singer, *The Life You Can Save* [New York: Random House, 2009], 15).

26. Aquinas, *On Law, Morality, and Politics*, *ST* II-II 66.7.

27. Ibid.

28. Ibid., II-II 66.8.

29. Ibid., I-II, 105.2.

30. John Locke, *The Second Treatise of Government*, ed. C. B. Macpherson (Indianapolis: Hackett, 1980), §6.

31. Ibid., §25.

32. C. B. Macpherson, introduction to *Second Treatise of Government*, by John Locke (Indianapolis: Hackett, 1980), xvii.

33. While Locke is not as forceful as he could be in underlining this ongoing obligation, this may be in part because of his optimism that the accumulation and rational investment of capital will lead to a shared prosperity that will generally alleviate extreme need. This leads some, like Steven Forde, to suggest that it is not that Locke thinks the right to private property trumps the right of necessity, but that capitalist economies built on private property protections will generally bring an end to unmet necessity. See Steven Forde, "The Charitable John Locke," *The Review of Politics* 71 (2009): 428–58.

34. John Locke, *The First Treatise of Government*, in *Two Treatises of Government*, ed. Peter Laslett (Cambridge: Cambridge University Press, 1988), §42. Though Locke appeals to religion here and in a variety of places in the *Second Treatise*, he does not seem to think that his political work depends on religious commitments. This is most clear in the *Second Treatise*, in which his religiously based arguments are always backed up by rational argumentation.

35. John Locke, "Venditio," in *John Locke: Political Writings*, edited by David Wootton (Indianapolis: Hackett, 1993), 445.

36. John Locke, "Draft of a Representation Containing a Scheme of Methods for the Employment of the Poor," in *John Locke: Political Writings*, ed. David Wootton (Indianapolis: Hackett, 1993).

37. Locke, *Second Treatise*, §135.

38. Emmanuel Levinas, "Secularization and Hunger," trans. Bettina Bergo, *Graduate Faculty Philosophy Journal* 20/2–21/1 (1998): 10.

39. Emmanuel Levinas, *Totality and Infinity*, trans. Alphonso Lingis (Pittsburgh: Duquesne University Press, 1969), 127.

40. Francois Rabelais, *Pantagruel, Book 4–5*, trans. William Francis Smith (London: Watt, 1893), 231. Cited in Levinas, "Secularization and Hunger," 8.

41. Levinas, "Secularization and Hunger," 11.

42. Ibid., 10.

43. Emmanuel Levinas, *God, Death, and Time*, trans. Bettina Bergo (Stanford, CA: Stanford University Press, 2000), 170.

44. Levinas, *Totality and Infinity*, 201.

45. Emmanuel Levinas, *Nine Talmudic Readings*, trans. Annette Aronowicz (Bloomington, IN: Indiana University Press, 1990), 133.

46. Emmanuel Levinas, *Otherwise than Being or Beyond Essence*, trans. Alphonso Lingis (Pittsburgh: Duquesne University Press, 1998), 56.

47. Levinas, *Nine Talmudic Readings*, 133.

48. For an optimistic analysis of such programs see Joseph Hanlon, Armando Barrientos, and David Hulme, *Just Give Money to the Poor: The Development Revolution from the Global South* (Sterling, VA: Kumarian Press, 2010).

Chapter 3

Food Deserts and Lockean Property

J. M. Dieterle

The expression "food desert" was first used during a 1990s study in Glasgow by the Low Income Project Team of the Nutrition Task Force of Great Britain.[1] The team described food deserts as "areas of relative exclusion where people experience physical and economic barriers to accessing healthy food."[2] Distance is the primary physical barrier: access to healthy food is limited because the nearest grocer is too far away to be easily accessible. Food deserts have been identified in many developed countries, including the United Kingdom,[3] Australia,[4] Canada,[5] and New Zealand.[6] However, the most extensive research on and mapping of food deserts has occurred in the United States.[7]

The U.S. Department of Agriculture (USDA) classifies an urban area as a food desert if at least 20 percent of the population lives below the poverty level and there is no mainstream grocery store selling fresh and nutritious food within one mile. Rural areas are classified as food deserts if the nearest mainstream grocer is more than ten miles away.[8] But it is important to note that the measurable distance does not tell the whole story. As Hillary J. Shaw notes in *The Consuming Geographies of Food*, one's physical distance from a grocer, if one must walk, is mediated by multiple factors.[9] Walking to the nearest grocer might involve climbing a steep hill or crossing one or more busy roads. There may be no sidewalk or pedestrian path along the most direct route. Heavy grocery bags can impact one's ability to walk long distances to purchase food, especially if one is disabled. The route to the grocer may be through an area that is unsafe at certain times of the day. In rural areas, public transportation may not be available or, if available, may not be reliable. Those who do not have a car or cannot drive may be unable to reach a grocer that stocks healthy food options, even if there is such a grocer within the ten-mile radius. The USDA classifications thus may underestimate the number of people who do not have ready access to healthy, affordable food. Nonetheless, using USDA criteria, 23.5 million U.S. residents live in food deserts.[10]

The demographics of food deserts are telling. Most food deserts in the United States are found in areas that are predominantly nonwhite.[11] In fact, according to the USDA, in all census tracts other than those that are urban and very dense, "the higher the percentage of minority population, the more likely the area is to be a food desert."[12] This is true of both urban and rural areas. The proportion of minorities in urban food deserts is 53 percent higher than in urban nonfood deserts. In rural areas, the minority population is 65 percent higher in food desert tracts than in nonfood desert tracts.[13] Poverty levels and unemployment rates are higher in food desert tracts, while median family incomes are substantially lower.[14] However, a 2013 study determined that neighborhoods in the United States that are predominantly African American suffer from the most limited access to venues that sell fresh, healthy food, *regardless of income*. When comparing census tracts of equal poverty levels, black neighborhoods had the fewest supermarkets and white neighborhoods had the most. The authors of the study conclude that poverty level and race are independent indicators of supermarket availability, and that poor African American neighborhoods face a "double jeopardy."[15]

The World Health Organization lists three facets of food security: (1) food availability (that a sufficient quantity of healthy, nutritious food is available on a consistent basis); (2) food access (that one has sufficient resources to access healthy, nutritious food); and (3) food use (that one has the knowledge of basic nutrition and sanitation to properly feed oneself).[16] Here, I wish to focus on facet (1): food availability. When there are physical barriers to accessing healthy food, food is essentially unavailable. Food insecurity is the direct result. Importantly, unhealthy food *is* often readily available in areas classified as food deserts. Fast food restaurants and "fringe" food establishments (gas stations, liquor stores, party stores, dollar stores, convenience stores, etc.) are common in food deserts.[17] Such establishments typically carry only a small selection of processed food products. These food products tend to be unhealthy: they are almost always high in salt, fat and sugar and have limited nutritional value.[18] Thus, the "food" available in food deserts is not sufficient for food security.

My goal in this chapter is to argue that food deserts are a serious matter of justice. This may seem uncontroversial. And, in fact, on most philosophical accounts of justice, it is. On a Rawlsian theory of justice, for example, a sketch of the argument would go something like this: one cannot participate as a free and equal citizen when one is food insecure.[19] Access to nutritious food is a precondition for the attainment of all primary goods, and food distribution should be subject to the difference principle.

The distribution of food ought to benefit everyone, and any inequality in its distribution ought to be to the benefit of the least well off. But, clearly, the distribution of food does *not* benefit everyone. Food insecurity is a direct result of food deserts. Further, food deserts harm the least well off disproportionately. More than half of the 23.5 million U.S. residents living in food deserts are in a low-income category.[20] Likewise, food deserts identified in the UK, EU, Australia, Canada and New Zealand are in low-income areas.[21] Clearly, then, the current distribution of food does *not* benefit the least well off, but, instead, makes life more difficult for the least well off. So the argument for injustice of food deserts is fairly straightforward from the Rawlsian perspective.

The argument is also fairly clear-cut from a utilitarian theory of justice. The net balance of social utility that results from the current distribution of food is negative. Not only are the inhabitants of food deserts harmed by the current distribution, but, given the health outcomes associated with food deserts and food insecurity, the consequences for public health are quite dire.[22] Direct costs of food deserts are borne by the individuals who reside in such communities, given that the quality of available food has a profound and negative impact on their health. Indirect costs are borne by the health care industry (who must treat diseases that result from poor diets), by employers (through lost productivity), and by taxpayers and governmental agencies (who in some cases shoulder the financial burden of the health effects of food deserts).[23] So the argument for the injustice of food deserts is also fairly straightforward from the utilitarian perspective.

The interesting question is how a libertarian theory of justice would assess food deserts. One might think that a libertarian would find food deserts unfortunate, but ultimately not a matter of justice. After all, they are the direct outcome of private property rights and capitalistic free markets. Grocery stores went where the money was: to the suburbs.[24] However, I argue that a plausible interpretation of the Lockean theory of property entails the injustice of food deserts. Thus, one who holds a theory of justice that depends on or substantially includes a Lockean theory of property has reason to support limitations on property rights to address the problem of food access. I begin with Locke himself and discuss why he should be committed to the injustice of food deserts. I then turn to Robert Nozick's historical theory of justice and argue that even he should admit that food deserts warrant property rights limitations.

I. LOCKE

In a Lockean state of nature, the earth and its resources are owned in common by all individuals. Further, individuals have natural rights to preservation and to preserve themselves. Gopal Sreenivasan distinguishes between these two rights as follows:

> These two natural rights may be distinguished by considering the particular duty each right imposes on others when possessed by a rights-bearer. In the former case, others have a duty to refrain from directly endangering the life of the rights-bearer; in the latter case, others have a duty to refrain from impeding the rights-bearer from actively preserving herself.[25]

Since we have a right to preserve ourselves, we may take from the commons what we need to exercise that right. Suppose, for example, there is a clump of blackberry bushes in my vicinity. I may pick and eat blackberries from those bushes to nourish myself. Further, if I pick more than I can eat, I may keep the surplus as my own. Locke's theory allows us to appropriate, as our own property, any surplus that arises from our labor.

Property acquisition is not unlimited, though. Indeed, Lockean property acquisition is subject to three important limitations. (1) The spoilage limitation: one cannot appropriate more than one can use before it spoils.[26] If I pick more blackberries than I can consume before they rot, I am not entitled to keep the entire surplus as my own. (2) The sufficiency condition (often called the "enough and as good" proviso): one must leave "enough and as good" for others.[27] On the (implausible) assumption that blackberries are the only means of fulfilling one's dietary needs in the traversable area, I am not entitled to take all of them from the commons. To appropriate the entire surplus as my own would be to leave others in danger of starvation and thus actively impede them from preserving themselves. Since the blackberries are owned in common by all, I must be sure to leave enough for others to eat. (3) The charity limitation: individuals who are incapable of laboring to produce their own means of sustenance are entitled to subsistence levels of charity from others, provided those others have a surplus.[28] So the charity limitation limits individual acquisition when others are in need and incapable of laboring to support themselves.

Land appropriation is legitimated in much the same way as appropriation of perishable items growing on land owned in common. If I cultivate and improve a plot of land through my labor, then I am entitled to call that land my own, provided I have respected the Lockean limitations on appropriation.

The important Lockean limitation, for my purposes, is (2): the sufficiency condition. In *The Limits of Lockean Rights in Property,* Gopal Sreenivasan argues that the sufficiency condition is Locke's answer to "the consent problem." Locke's goal in Chapter V of *The Second Treatise of Government* is to legitimate individual appropriation of land and resources. Given the assumption that the earth and its resources are originally owned in common, individual appropriation seems to be a kind of theft. If one could obtain the consent of the co-owners, of course, appropriation would be legitimate. But it is practically impossible to secure the consent of each and every co-owner when the earth and its resources are commonly owned by all. However, if enough and as good remains once one has appropriated that of which one can make use, then no one is harmed by the appropriation and no one's right to self-preservation is transgressed. As such, consent can be foregone.

The sufficiency condition safeguards individuals' rights to preserve themselves and thus entails that a legitimate appropriation may not jeopardize said individuals' access to those resources necessary to produce their sustenance.[29] Importantly, an appropriation *would* cause harm, and thus *would* violate the sufficiency condition if said appropriation were to impede a fellow rights-bearer from exercising her right to preserve herself. The sufficiency condition ensures that a property appropriation is illegitimate if and when such harm occurs. In a time and place where arable land is abundant, violations of the sufficiency condition rarely, if ever, occur. There's plenty of land to go around, so no one is impaired in exercising her right to preserve herself. However, land abundance is likely to end with the advent of civil society and the introduction of money. Since money is imperishable, one can amass large amounts of property without the threat of spoilage. Once individuals are free to accumulate such imperishable property, resource scarcity is a likely result. In turn, resource scarcity brings with it the potential to impair others' access to the materials needed to produce their sustenance and thus impede their exercise of the right to self-preservation. But, importantly, given the sufficiency condition, access to the means of preservation is protected. Individual acquisitions that would endanger others' access to said means continue to be illegitimate.

It is worth noting that the claim that the sufficiency condition continues to limit individual appropriation in civil society is not universally accepted by Locke scholars. On the alternate reading of Locke, *both* the spoilage limitation and the sufficiency condition are discharged with the introduction of money and the advent of civil society.[30] However, if the function of the sufficiency condition is to safeguard the right to preserve oneself, then there is every reason to believe that Locke intended it to remain in place.[31] For Locke, the right to preserve oneself is a natural right which cannot be superseded by civil society. The landless cannot be denied access to the means of preservation, even when land itself is scarce. The sufficiency condition protects this access. Note, though, that there is more than one way for the sufficiency condition to be met. Locke does not require that enough and as good *of the same kind* remain to legitimate an appropriation. As long as others have access to the means of preservation, the sufficiency condition is satisfied. Hence, if landowners employ as many people as the land could have sustained had it not been appropriated, then employed but landless individuals presumably are able to earn wages and exchange those wages for sustenance. Their right to preserve themselves is thus theoretically not violated.[32]

In practice, however, things are not so clear. We live in a post-agrarian society. As such, most of us exercise our right to preserve ourselves by laboring for someone else. We labor for wages and exchange our wages for sustenance. However, those living in food deserts are *unable* to exchange their wages for sustenance. Since there is no readily accessible venue for purchasing affordable, nutritious food, food deserts leave individuals without the means to procure the basic necessities of a productive life. As such, residents of food deserts cannot secure the material preconditions of the right to preserve themselves. Enough and as good is *not* available to them: because of private property rights and capitalistic free markets, they are food insecure. Note that this is so *even if* they labor for wages that would be sufficient to purchase their own sustenance if nutritious, healthy food were available.

Of course, there is a difference between *bare subsistence levels* of food and food *adequate to preserve a productive, healthy life*. My argument is that food deserts violate the sufficiency condition because their residents are denied access to the basic necessities of a productive life. But one might contend that the sufficiency condition does not require that level of food security; that Lockean property theory entitles the residents only to bare subsistence levels of food. On this alternative interpretation, one could argue that food deserts are, in fact, consistent with the sufficiency condition, since a subsistence level of food *is* available in food deserts. After all, one *can* survive (at least provisionally) on processed, unhealthy food. However, this alternative interpretation takes a very narrow reading of the right to preserve oneself. My contention is that it is too narrow.

Recall that the right to preserve oneself gives others correlative duties; others may not impede the right-bearer from actively preserving herself. Speaking of these correlative duties, Locke says:

> Every one as he is bound to preserve himself, and not to quit his station wilfully; so by the like reason when his own Preservation comes not in competition, ought he, as much as he can, to preserve the rest of Mankind, and may not unless it be to do Justice on an Offender, take away, or impair the life, or what tends to the Preservation of the Life, Liberty, Health, Limb, or Goods of another.[33]

As Locke notes, our health is essential to the preservation of our life. One may be able to provisionally survive on processed, unhealthy food, but one could not preserve one's life over the long term. So we should read the right to preserve ourselves more inclusively, as including the right to actively preserve our *health.* Thus, if we cannot access the means to procure the basic necessities of a productive life—those things which will keep us healthy and productive—we are thereby impeded in exercising our right to preserve ourselves. As such, when property acquisitions or transfers thwart an individual's exercise of her right to preserve her *health*, they equally thwart her exercise of the right to preserve her life, and the sufficiency condition is violated.[34] A bare subsistence level of unhealthy food is not *enough*, nor is it *as good* as nutritious, healthy food.

Those who are food insecure because they lack access to healthy food are substantially harmed by property acquisitions and transfers that limit food access. Recall that fast food restaurants and "fringe" food establishments (gas stations, liquor stores, party stores, dollar stores, convenience stores, etc.) are common in areas classified as food deserts. The food for sale in such establishments is typically highly processed and high in fat, sugar and salt. When there are physical barriers to accessing healthy food, but unhealthy food is readily available, negative health outcomes are the likely result. And, in fact, that is exactly what we find. Residents of food deserts have higher rates of diabetes, cancer, obesity and heart disease than residents of other areas. Note that these results are independent of other contributing factors such as income, race and education.[35] A study by the Mari Gallagher Research and Consulting Group found that the death rate from diabetes for those areas of Chicago that are the most out of balance is more than double that of other communities in the surrounding area.[36] The same study found that obesity rates increase the further one gets from a grocery store.[37] And, of course, obesity is a contributing factor in many chronic diseases: heart disease, stroke, diabetes, high blood pressure, some forms of cancer and others.

Food insecurity has devastating effects on children. Studies have shown that food insecure children are twice as likely as food secure children to have fair or poor health. Hospitalization rates in food insecure children are roughly one-third greater than those for children living in food security.[38] Food insecure children have more frequent stomachaches, headaches and colds than children in food secure households.[39] Anxiety scores are more than double those of food secure children.[40] Food insecurity is also linked to developmental problems. Academic impairment is evident in both reading and math scores for children who were food insecure in kindergarten.[41] Food insecure children are more likely than food secure children to have repeated a grade.[42] Iron-deficiency anemia, often associated with food insecurity, impacts children's development of basic motor and social skills.[43] Alaimo, Olson and Frongillo found that children who had suffered from and were treated for iron-deficiency anemia as infants still suffered from impaired memory and social functioning more than ten years later.[44] Food insecure teenagers have difficulty getting along with other children and are more likely to be suspended from school.[45] Food insecure adolescents (fifteen to sixteen years old) are more likely to have depressive disorders and suicide symptoms.[46] And so on.

It seems clear that enough and as good is not available to those who live in food deserts. "Food" is available, but not the kind of food that will allow one to maintain a healthy, productive life. The Lockean sufficiency condition is a limit on property rights. Since the natural right to preserve oneself is violated if individuals are denied access to the materials needed for their preservation, and food deserts deny people access to such basic materials, property acquisitions and transfers that result in food deserts are in violation of the sufficiency condition. As such, they are unjust. Even though food deserts are the result of private property rights and capitalistic markets, a libertarian in the Lockean tradition has good reason to support state intervention and limitations on property rights to correct the injustice.[47]

Of course, there are assumptions behind Locke's argument that contemporary libertarians might not endorse, such as natural law and the original common ownership of the earth and its resources. I thus now turn to a contemporary libertarian, to see how the argument would play out without these assumptions.

II. NOZICK

In Nozick's original state, the earth and its resources are *unowned*—not owned in common as in Locke's state of nature. Recall that the sufficiency condition is necessary for Locke as a mechanism for legitimating appropriation without the consent of the co-owners. But since nothing is owned in Nozick's original state, the consent problem does not arise. Nevertheless, Nozick recognizes that property acquisition should be limited by a theory of justice. His account of justice includes three principles: (1) a principle delineating the process for justly acquiring unowned things (the principle of justice in acquisition); (2) a principle that governs the transfer of property from one person to another (the principle of justice in transfer); and (3) a principle that spells out the process for rectifying past injustices (the principle of rectification). The principle of justice in acquisition includes a version of the sufficiency condition. Nozick requires that "the situation of others is not worsened"[48] by an appropriation, in the sense that the others are no longer "able to use freely (without appropriation) what [they] previously could."[49] He says:

> I assume that any adequate theory of justice in acquisition will contain a proviso similar to the weaker of the ones we have attributed to Locke. A process normally giving rise to a permanent bequeathable property right in a previously unowned thing will not do so if the position of others no longer at liberty to use that thing is thereby worsened.[50]

The sufficiency condition also constrains later actions. Nozick continues:

> Each owner's title to his holding includes the historical shadow of the Lockean proviso on appropriation. This excludes his transferring it into an agglomeration that does violate the Lockean proviso and excludes his using it in a way, in coordination with others or independently of them, so as to violate the proviso by making the situation of others worse than their baseline situation. Once it is known that someone's ownership runs afoul of the Lockean proviso, there are stringent limits on what he may do with (what it is difficult to any longer unreservedly call) "his property."[51]

The principle of justice in transfer is thus also subject to the sufficiency condition. One cannot transfer property to another if said transfer makes it the case that others are worse off. Of course, the key question is: worse off than what? Nozick takes the baseline to be a system with *no private property whatsoever*, and he argues that unregulated free markets in a liberal property regime will ensure that the sufficiency condition is met.[52]

Nozick's example of an unjust use of property has to do with water. He notes that an individual may not appropriate or purchase all of the access rights to drinkable water in a given location and then charge exorbitant prices for it.[53] Nor, presumably, could the individual decide to lease all of her access rights to a business that bottles and sells water. Such actions would be in violation of the sufficiency condition since the inhabitants of the area would not have access to one of the things necessary to sustain their lives. They would thus be much worse off than they would be in a no-property situation. As a result, the property rights of the individual owner would be limited.[54]

The sufficiency condition limits an individual's property rights when others are made worse off than their baseline situation. The constraints that limit property rights in such situations also limit property rights when *multiple* owners are involved. Suppose that many individuals independently acquire rights of access to drinkable water in a particular area and none wish to provide water to the local community. (Perhaps all of them wish to profit from leasing their rights to businesses that bottle water and then sell it at exorbitant prices.) Suppose further that all water access points are exhausted by the multiple acquisitions and, as a result, there is no way for local residents to obtain drinkable water without infringing an owner's property rights. In such a situation, the residents are left without access to potable water; they are in the same situation as they would be if one individual owned all of the water rights and denied them access. Said inhabitants are much worse off than they would be in a no-property situation—enough and as good is not available to them—and the sufficiency condition is violated. Here, too, the property rights of the owners would be limited by Nozick's version of the sufficiency condition.

Nozick argues that such situations will rarely—if ever—occur in a system with private property and unregulated free markets:

> I believe that the free operation of a market system will not actually run afoul of the Lockean proviso. . . . If this is correct, the proviso will not play a very important role in the activities of protective agencies and will not provide a significant opportunity for future state action.[55]

Nozick's argument for the claim that a free market system in a liberal property regime will ensure that the sufficiency condition is met relies on the productive capacity of private property. He argues that we are *all* better off than we would be in a no-property situation. The system of private property

> increases the social product by putting means of production in the hands of those who can use them most efficiently (profitably); experimentation is encouraged, because with separate persons controlling resources, there is no one person or small group whom someone with a new idea must convince to try it out; private property enables people to decide on the pattern and types of risks they wish to bear, leading to specialized types of risk bearing; private property protects future persons by leading some to hold back resources from current consumption for future markets; it provides alternate sources of employment for unpopular persons who don't have to convince any one person or small group to hire them, and so on.[56]

For the sake of argument, I will assume that Nozick is right about the overall benefits of a system of private property. However, note that the benefits do not necessarily accrue to everyone. Those who are without land or money reap few rewards from a system that restricts their access to resources that were originally unowned. The intent of the sufficiency condition, I take it, is to protect *individuals* from being harmed by others' appropriations and transfers of property.[57] That is, the intent is not to ensure that society as a whole is no worse off than it would be without a system of private property, nor even that the majority is no worse off, but, instead, its intent is to make sure *no individual* is made worse off by an appropriation or transfer than she would be in a no-property system. By this standard, it is not clear how a system of unregulated markets is supposed to guarantee that the sufficiency condition is met.

Given that residents of food deserts lack access to the very means of preservation, we seem to have a case where Nozick's sufficiency condition is, in fact, violated. At least prima facie, residents seem to be much worse off than they would be in a system without private property.[58] In fact, the property transfers that have resulted in food deserts are directly analogous to the property transfers that result in individuals being left without access to water in Nozick's own example. Inhabitants of food deserts are food insecure because they lack access to nutritious, affordable food—something necessary for a healthy, productive life. Furthermore, that this is so is *a direct result of* free markets and private property. Grocers have moved to more affluent locations, leaving residents without access to nutritious, healthy food. Of course, Nozick's sufficiency condition is not intended as an end-state principle.[59] He says, "It focuses on a particular way that appropriative actions affect others, and not on the structure of the situation that results."[60] But food deserts emerged as a direct result of appropriative actions and transfers of property. Grocers moved to the suburbs, convenience stores and liquor stores took the place of neighborhood markets, and so on. Access to the means of preservation has been limited directly by transfers of property. As such, the sufficiency condition should be in effect. It should limit transfers of property when said transfers leave individuals without access to nutritious, affordable food.

Enough and as good is *not* available to residents of food deserts. Hence, food deserts are in violation of Nozick's version of the sufficiency condition. They are thus unjust, even in a Nozickian minimal state, and the principle of rectification should be invoked to correct the injustice.

III. CONCLUSION

I have argued that food deserts are a serious matter of justice. My primary thesis is that even someone committed to libertarian principles and a Lockean theory of property has reason to support limitations on property rights to address the problem of food access in food deserts. Note, though, that there is much more to be said about the injustices surrounding food deserts. Improving physical access will not, alone, eliminate food insecurity in food-rich nations.[61] As Hillary J. Shaw notes in *The Consuming Geographies of Food,* the problem is multilayered. Even were the problem of physical access solved, a financial layer would remain. Access means little if people cannot afford to purchase fresh, healthy food. Further, there is a third layer, which is more complicated and perhaps less visible: in some cases, food preferences are the problem. People who have access to and the means to purchase fresh, healthy food may very well choose not to eat it.[62] Finally, note that food deserts are an issue of food distribution, but, as so many political theorists have stressed, distribution alone is never the whole story.[63] We need to address the causes of food deserts (poverty, oppression) before we can really solve the problem.

Nonetheless, improving physical access to healthy food is an important first step. Currently, there are grassroots efforts to improve the quality of food in food deserts. Such efforts include urban farming, community gardens, expanded farmers' markets in urban environments, and food trucks in underprivileged areas.[64] But much more is called for to fully address the problem of physical access to fresh, healthy food. We need federal, state and local mobilization to effect change. For example, zoning ordinances can be implemented to limit property rights when exercise of them would result in food deserts and state finances can be utilized to develop food retail in areas where there is little to no access to fresh food.

Pennsylvania's Fresh Food Financing Initiative (FFFI) is one example of a governmental effort to address the food desert problem. The FFFI is a grants and loan program that was set up to encourage development of fresh food retail in underserved areas. Pennsylvania contributed $30 million in state funds to the initiative. So far, eighty-eight projects have been funded and food access has improved for roughly 400,000 residents of Pennsylvania because of FFFI.[65] Due in part to the success in Pennsylvania, the federal government launched the Healthy Food Funding Imitative (HFFI), a partnership between the U.S. Departments of Treasury, Agriculture and Health that aims to improve food access in underserved areas. Like the FFFI, the HFFI provides grants and loans to develop fresh food retail in areas that are now classified as food deserts.[66] Yet, still, this is not enough. Food access remains problematic for millions of people in the United States and other food-rich developed countries.

I conclude with a quote from George A. Kaplan, Thomas Francis Collegiate Professor of Public Health at the University of Michigan. Commenting on a report by the Mari Gallagher Research and Consulting Group on food access in Chicago, he says:

> I find the use of the term "food desert" particularly interesting. A desert is, of course, a place distinguished by the absence of vegetation, rain, etc., which is the sense in which the word is used in this report. Food deserts are defined as "areas with no or distant grocery stores." But the word "desert" is also a verb—"to leave someone without help or in a difficult situation and not come back." This seems to me to capture an important dimension of food deserts not conveyed by the noun. The verb "desert" focuses on action and agency, emphasizing that the lack of access to good food in some areas is not a natural, accidental phenomenon but is instead the result of decisions made at multiple levels by multiple actors. By focusing on this latter meaning, we can find room for changes to be effected, for different decisions to be made in the future, for movement toward actions that can improve access to healthy food for those who have been deserted. In doing so, we can help in at least one way to improve uneven opportunities, and perhaps provide better health as well.[67]

NOTES

1. See Neil Wrigley, "Food Deserts in British Cities: Policy Context and Research Priorities," *Urban Studies* 39 (2002): 2030 and Hillary J. Shaw, *The Consuming Geographies of Food* (New York: Routledge, 2014), 105.

2. Vmt Reisig and A. Hobbiss, "Food Deserts and How to Tackle Them: A Study of One City's Approach," *Health Education Journal* 59 (2000): 138.

3. See Shaw, *Consuming Geographies*, 111–15 for a discussion of food desert research in the UK. Shaw notes that there is no standard metric for classifying an area as a food desert in the UK. Distance from a grocer ranges from ¼ mile to 1 mile in the literature. See also Wrigley, "Food Deserts in British Cities" for a discussion of the controversy over whether there actually are food deserts in the UK.

4. See Kylie Ball, Anna Timperio, and David Crawford, "Neighborhood Socio-economic Inequalities in Food Access and Affordability," *Health & Place* 15 (2009): 578–85, for a discussion of food deserts in Melbourne, Australia.

5. See Kristian Larsen and Jason Gilliland, "Mapping the Evolution of 'Food Deserts' in a Canadian City: Supermarket Accessibility in London, Ontario, 1961–2005," *International Journal of Health Geographics* 7 (2008): 16.

6. See J. Pearce, T. Blakely, K. Witten, and P. Bartie, "Neighborhood Deprivation and Access to Fast-Food Retailing: A National Study," *American Journal of Preventative Medicine* 32 (2007): 375–82 for discussion of New Zealand food deserts.

7. Although several local studies have been carried out in the UK and other developed nations, as yet, the United States is the only place where systematic mapping has taken place. Furthermore, "A recent systematic review of 48 studies from 1966 through 2007 . . . shows equivocal findings about the existence of food deserts in many European countries—but clear evidence of disparities in food access in the United states by income and race." Paula Tarnapol Whitacre, Peggy Tsai, and Janet Mulligan, "The Public Health Effects of Food Deserts: Workshop Summary" (Washington, D.C.: National Academy of Sciences, 2009), 30. The review cited is Julie Beaulac, Elizabeth Kristjansson, and Steven Cummins, "A Systematic Review of Food Deserts, 1966–2007," *Preventing Chronic Disease* 6 (2009): A105.

8. USDA, "Food Deserts," accessed June 3, 2013, http://apps.ams.usda.gov/fooddeserts/foodDeserts.aspx.

9 See Shaw, *Consuming Geographies*, 110–11.

10. USDA, "Food Deserts."

11. See Paula Dutko, Michele Ver Ploeg, and Tracey Farrigan, "Characteristics and Influential Factors of Food Deserts," USDA Economic Research Report 140 (2012): 11–13; Lisa M. Powell, Sandy Slater, Donka Mirtcheva, Yanjun Bao, and Frank J. Chaloupka, "Food Store Availability and Neighborhood Characteristics in the United States," *Preventative Medicine* 44 (2007): 189–95; and K. Morland, S. Wing, A. Diez Roux, and C. Poole, "Neighborhood Characteristics Associated with the Location of Food Stores and Food Service Places," *American Journal of Preventative Medicine* 22 (2002): 23–9.

12. Dutko, ver Ploeg, and Farrigan, "Characteristics," iii.

13. Ibid., 11.

14. Ibid., 13. This holds in both urban and rural areas.

15. Kelly M. Bower, Roland J. Thorpe Jr., Charles Rohde, and Darrell J. Gaskin, "The Intersection of Neighborhood Racial Segregation, Poverty, and Urbanicity and its Impact on Food Store Availability in the United States," *Preventative Medicine* 58 (2014): 33–9.

16. World Health Organization (WHO), "Food Security," accessed July 15, 2014, http://www.who.int/trade/glossary/story028/en/. The FAO adds a fourth: stability (that access to food is stable over time and not subject to shocks, economic or otherwise). The WHO criteria fold stability into facet (1) by including "on a consistent basis." See FAO, "Food Security: Policy Brief" (Rome: Food and Agricultural Organization of the United Nations, 2006).

17. The locution "fringe food locations" originates with the Mari Gallagher Research and Consulting Group. For their research on food deserts, see http://marigallagher.com/.

18. See Mari Gallagher, "Examining the Impact of Food Deserts on Public Health in Detroit" Mari Gallagher Research Group, 2007.

19. A complete assessment of food deserts from the Rawlsian perspective would require a full defense of this claim. The idea is that when one lives in food insecurity, one's energy and resources must be devoted to self-preservation. Under such conditions, it would be difficult to assert one's rights, participate in civil society, enjoy freedom of movement, have free choice among occupations, etc. In short, one could not pursue one's conception of the good life if one is living in food insecurity.

20. USDA, "Food Deserts."

21. See Shaw, *Consuming Geographies*; Ball, Timperio, and Crawford, "Neighborhood Socioeconomic Inequalities"; Larsen and Gilliland, "Mapping the Evolution"; and Pearce, Blakely, Witten, and Bartie, "Neighborhood Deprivation."

22 See below for a discussion of the health outcomes related to food deserts.

23 Mari Gallagher, "Examining the Impact of Food Deserts on Public Health in Chicago" Mari Gallagher Research and Consulting Group, 2006, 10.

24. The evolution of food deserts is actually slightly more complicated than this. Causes include changes in housing, transportation, and banking policies and changes internal to the grocery industry. See Allison Karpyn and Sarah Treuhaft, "The Grocery Gap: Finding Healthy Food in America," in *A Place at the Table*, ed. Peter Pringle (New York: Public Affairs, 2013), for discussion. Even so, they are the direct result of private property rights and capitalistic markets.

25. Gopal Sreenivasan, *The Limit of Lockean Rights in Property* (New York: Oxford University Press, 1995), 23–4.

26. "But how far has he given it us? *To* enjoy, As much as any one can make use of to any advantage of life before it spoils . . ." John Locke, *Two Treatises of Government*, introduction and notes by Peter Laslett (New York: Cambridge University Press, 1963), 2nd Treatise, V, 31, 9.

27 "Nor was this *appropriation* of any parcel of *Land*, by improving it, any prejudice to any other Man, since there was still enough, and as good left . . ." Ibid., 2nd Treatise, V, 33, 1.

28 "As *Justice* gives every Man a Title to the product of his honest Industry, and the fair Acquisitions of his Ancestors descended to him; so *Charity* gives every Man

a Title to so much out of another's Plenty, as will keep him from extreme want, where he has no means to subsist otherwise . . ." Ibid., 1st Treatise, IV, 43, 35.

29. Sreenivasan, *Limits*, 5.

30. Those who think the sufficiency condition remains in force after the advent of civil society include Robert Nozick, James Tully, A. John Simmons, and Gopal Sreenivasan. Those who think it does not include C. B. Macpherson and Jeremy Waldron. Waldron argues that the sufficiency condition is a sufficient (not necessary) condition on appropriation, even in the state of nature. See Robert Nozick, *Anarchy, State, and Utopia* (New York: Basic Books, 1974); James Tully, *A Discourse on Property: John Locke and His Adversaries* (Cambridge: Cambridge University Press, 1980); A. John Simmons, *The Lockean Theory of Rights* (Princeton: Princeton University Press, 1992); Sreenivasan, *Limits*; C. B. Macpherson, *The Political Theory of Possessive Individualism: Hobbes to Locke* (Oxford: Clarendon Press, 1962); and Jeremy Waldron, *The Right to Private Property* (Oxford: Clarendon Press, 1988).

31. Furthermore, note that although Locke explicitly suspends the spoilage limitation when money is introduced, he does not so suspend the sufficiency condition. See Locke, *Two Treatises*, 2nd Treatise, V, 47–50.

32. It is worth noting that Sreenivasan argues that Locke doesn't take the sufficiency condition seriously enough, that, in fact, it entails that appropriation is legitimate up to only the largest universalizable share. But the injustice of food deserts will follow regardless of whether we follow Sreenivasan this far. So here I ignore his contention and concentrate on the more limited Lockean view. After all, if the injustice follows on the limited view, it is more compelling. See Sreenivasan, *Limits*.

33 Locke, *Two Treatises*, 2nd Treatise, II, 6.

34. I assume here the sufficiency condition governs property transfers as well as original acquisition. In this, I follow Nozick and Sreenivasan. See Nozick, *Anarchy, State and Utopia,* and Sreenivasan, *Limits*.

35. See Mari Gallagher, "Food Desert and Food Balance Community Fact Sheet" Mari Gallagher Research and Consulting Group, 2010, accessed June 1, 2013, http://www.fooddesert.net/wp-content/themes/cleanr/images/FoodDesertFactSheet-revised.pdf.

36. Gallagher, *Chicago*, 7. Note that the Mari Gallagher Research and Consulting Group uses a different measure than does the USDA. Instead of measuring distance from a grocer, they use a metric that determines "food balance." The Food Balance Score of a particular area is determined by the distance to the nearest grocer (or other venue where one can purchase fresh, healthy food) divided by the distance to the closest fringe food location. The idea behind the Food Balance Score is that it indicates the level of difficulty in choosing a healthy food venue over a fringe food location.

37. Ibid., 9.

38. John T. Cook, Deborah A. Frank, Carol Berkowitz, Maureen M. Black, Patrick H. Casey, Diana B. Cutts, Alan F. Meyers, Nieves Zaldivar, Anne Skalicky, Suzette Levenson, Tim Heeren, and Mark Nord, "Food Insecurity is Associated with Adverse Health Outcomes Among Human Infants and Toddlers," *The Journal of Nutrition* 134 (2004): 1432–38. Note that this study controlled for confounding factors, including poverty.

39. Katherine Alaimo, Christine M. Olson, and Edward A. Frongillo, Jr., "Food Insufficiency and American School-Aged Children's Cognitive, Academic, and Psychosocial Development," *Pediatrics* 108 (2001): 44–53. This study controlled for confounding factors.

40. Linida Weinreb, Cheryl Wehler, Jennifer Perloff, Richard Scott, David Hosmer, Linda Sagor, and Craig Gundersen, "Hunger: Its Impact on Children's Health and Mental Health," *Pediatrics* 110 (2002): e41. This study controlled for confounding factors (housing status, mother's distress, and stressful life events).

41. Diana F. Jyoti, Edward A. Frongillo, and Sonya J. Jones, "Food Insecurity Affects School Children's Academic Performance, Weight Gain, and Social Skills," *Journal of Nutrition* 135 (2005): 2831–39. Developmental outcomes in this study were measured both with and without controls for confounding factors. The results cited are from the models with controls.

42. Alaimo, Olson, and Frongillo, "Food Insufficiency." This study controlled for confounding factors.

43. Ruth Rose-Jacobs, Maureen M. Black, Patrick H. Casey, John T. Cook, Diana B. Cutts, Mariana Chilton, Timothy Heeren, Suzette M. Levenson, Alan F. Meyers, and Deborah A. Frank, "Household Food Insecurity: Associations With At-Risk Infant and Toddler Development," *Pediatrics* 121 (2008): 65–72. This study controlled for confounding factors, including poverty.

44. Alaimo, Olson, and Frongillo, "Food Insufficiency." This study controlled for confounding factors.

45. Ibid.

46. Katherine Alaimo, Christine M. Olson, and Edward A. Frongillo, Jr., "Family Food Insufficiency, but Not Low Family Income, is Positively Associated with Dysthymia and Suicide Symptoms in Adolescents," *Journal of Nutrition* 132 (2002): 719–25. This study controlled for confounding factors.

47. See below for specific suggestions of limitations that might be a step in the direction of correcting the injustice.

48. Nozick, *Anarchy, State and Utopia*. 175.

49. Ibid., 176.

50. Ibid., 178.

51. Ibid., 180.

52. I address his argument for this claim below.

53. I've changed Nozick's example slightly, as his scenario takes place in a desert. The basic structure of the example is the same.

54. Nozick notes that the owner retains property rights over her water access, but that her rights are *limited* by the sufficiency condition. The rights are overridden to avoid catastrophe. See Nozick, *Anarchy, State and Utopia*, 180–81.

55 Nozick, *Anarchy, State and Utopia*, 182.

56. Ibid., 177.

57. "The very idea of a proviso is that it protects everyone." Karl Widerquist, "Lockean Theories of Property: Justifications for Unilateral Appropriation," *Public Reason* 2 (2010): 14.

58. Of course, we do not—and cannot—know what things would have been like if the earth's land and resources were unowned. However, it is clear that residents of food deserts are substantially harmed by the current system, in that they lack access to resources necessary for a productive life.

59. Lawrence Becker argues that there is no real difference between end-state theories and historical theories. Since all plausible historical theories will contain provisos limiting acquisitions and transfers, (a) they are just as substantive as end-state theories (the proviso contains the substance); and (b) they will require just as much "tinkering" to respect the proviso as end-state theories require to preserve the end-state (pattern). See Lawrence C. Becker, "Against the Supposed Difference Between Historical and End-State Theories," *Philosophical Studies* 41 (1982): 267–72.

60. Nozick, *Anarchy, State and Utopia*, 181.

61. Despite the fact that the United States is a food-rich nation, one out of every six Americans is food insecure. WHO, "Food Security."

62. Shaw, *Consuming Geographies*, 131.

63. See, for example, Iris Marion Young, *Justice and the Politics of Difference* (Princeton: Princeton University Press, 1990); and Elizabeth Anderson, "What is the Point of Equality?" *Ethics* 109 (1999): 287–337.

64. Note, however, that most of these efforts concentrate on urban environments. We need programs to alleviate food insecurity in rural food deserts, as well.

65. See The Food Trust, "What We Do," accessed November 16, 2014, http://thefoodtrust.org/what-we-do/supermarkets

66. See United States Department of Health and Human Services, "Community Economic Development Healthy Food Financing Initiative Projects," accessed February 24, 2015, https://www.acf.hhs.gov/hhsgrantsforecast/index.cfm?switch=grant.view&gff_grants_forecastInfoID=67242, for a description of the initiative.

67. Quoted in Gallagher, *Chicago*, 5.

Chapter 4

Food Deserts, Justice and the Distributive Paradigm

Jennifer Szende

This chapter rests on a widely shared intuition about social justice: that radically unequal distributions of material wealth are unfair and unjust.[1] In the context of food justice, we understand that inequality of distribution and access is a significant portion of the problem, and that the unfairness is part of the injustice. The relationship between unfairness, inequality and injustice takes up a significant portion of political philosophy's focus over the last century. Yet, there remains much to be learned by examining how these concepts function in real-world examples. This chapter examines the nature of food deserts in order to elucidate how unequal distributions come to constitute an injustice in the real world. My methodology follows Iris Marion Young's invitation to examine "justice" from a socially, historically and politically situated perspective.[2] Justice, even once elucidated, remains an abstract concept, but it ought nonetheless to be tested and examined with respect to its real-world applications and implications.

I use the example of food deserts to illustrate Young's argument. Food deserts show that some examples of injustice are most effectively defined and most readily diagnosed as distributive problems. In many respects, food deserts and other forms of environmental injustice are an illuminating example of, and indicator of, injustice understood in social power terms.[3] As a result, I present food deserts as an illuminating example of Young's theory: food deserts are partially constituted by inequality of distributions, but they are nonetheless the product of oppression and domination, powerlessness and corporate capital. Distributive analysis remains relevant as a diagnostic tool for identifying the injustice of food deserts, and ultimately redistribution serves as a necessary part of a remedy to the injustice. However, the injustice of food deserts is not wholly constituted by distributions, nor is redistribution sufficient to remedy the injustice of food deserts. In this chapter, food deserts and food justice serve to illustrate the need for a broad understanding of justice.

A food desert can be defined as the absence of large supermarkets or the absence of affordable healthy food in a particular neighborhood or community.[4] The problem is identified as more than mere misfortune because the absence defined by food deserts has been found to generally correlate with various forms of social hierarchies including marginalization along a variety of social vectors: race, gender, age, disability and minority status, to name a few. The city of Detroit as a whole is famously considered a food desert, but food deserts typically affect smaller geographic areas and populations, both rural and urban. In the United States, predominantly black neighborhoods have six times as many fast food restaurants and half as many supermarkets when compared with predominantly white neighborhoods.[5] In Canada, one study showed low-income areas to have half as many grocery stores and three times as many convenience stores as other areas.[6]

Given the health consequences of food insecurity, and the social power progenitors of food deserts, food deserts can be seen as both a cause and an effect of unjust social processes, or so I will argue. The central question in this chapter is how best to understand injustice in light of this example, and furthermore how food deserts can elucidate the underlying issue of the nature of injustice. At its heart, the very idea that access to nutritious food varies between populations is a distributive one, but food deserts are much more complex than mere inequality of access to material goods. Given that food deserts are widely accepted to form part of a story of broadly defined injustice, they are a particularly useful example for examining the concept of justice.

Furthermore, since we know that poor diet correlates with poor health and educational outcomes, food deserts are expected to have long-term implications for social inequality and power imbalances in these same communities.[7] Food deserts demonstrate some of the variety of forms that marginalization within affluent societies may take, and they do so both upstream and downstream of the oppression and domination that constitute injustice on Young's account. Oppression and domination have tangible distributive consequences, and unfair or unjust distributions produce and entrench existing oppression and domination.

Young and others have forcefully argued that there are a variety of reasons to mistrust definitions of justice that employ what she calls "the distributive paradigm."[8] That is, there are a variety of reasons to avoid simply identifying "injustice" with particular distributions of things.[9] To view justice and injustice as wholly constituted by distributions of material objects would suffer from reductionism.[10] Injustice is more profound than a mere lack of material objects or wealth; and injustice in the real world is more complex than distributive analysis alone suggests. Young understands injustice as institutional and structural, as involving power relations and hierarchies, and as entrenched and self-perpetuating in a variety of ways. And Young argues that the logic of distribution alone cannot adequately capture the injustice of unequal power relations and hierarchies.

Along with food deserts, a variety of injustices have a significant distributive aspect, and are readily identifiable in their distributive formulation, so an examination of the correlation between oppression and distribution can therefore be a useful diagnostic tool demonstrating the urgency of the justice claim.[11] In many such cases, redistribution will also serve as a necessary component of the response to injustice. However, given the complexity of the power relations that underlie unfair distributions, redistribution alone will not be sufficient as a just response—not even to unfair distributions themselves. Hence, this examination of the appropriate role for distribution in a theory of justice will conclude that distributions ought to serve both a diagnostic and a prescriptive role, but that distribution's role in justice ought not to be inflated.

I. YOUNG'S OBJECTION

The first task of this chapter will be to unpack Young's discussion of the distributive paradigm. In her *Justice and the Politics of Difference*, Young argues for a theory of justice that is less abstract, and better grounded in the practice of emancipatory social movements. Such a theory must account for distributions because "The immediate provision of basic material goods for people now suffering severe deprivation must be a first priority for any program that seeks to make the world more just."[12] Young argues that there are a variety of reasons to mistrust what she calls "the distributive paradigm," and in particular its definition of "social justice as the morally proper distribution of social benefits and burdens among society's members."[13] Young explains:

> Most theorists take it as given . . . that justice is about distributions. The paradigm assumes a single model for all analyses of justice: all situations in which justice is at issue are analogous to the situation of persons dividing a stock of goods and comparing the size of the portions individuals have. Such a model implicitly assumes that individuals or other agents lie as nodes, points in the social field, among whom larger or smaller bundles of social goods are assigned. . . . The distributive paradigm . . . implicitly assumes a social atomism, inasmuch as there is no internal relation among persons in society relevant to considerations of justice.[14]

The distributive paradigm is deployed to define justice in one of two ways, and both of them are problematic. According to Young's analysis of the distributive paradigm, either (1) justice is interpreted as reducible to facts about distributions of wealth and material objects or (2) rights and power relations are treated as objects that can metaphorically be distributed in just or unjust ways. In either case, the distributive paradigm misses the nonmetaphorical sense in which real power relations and social hierarchies are enacted. Real power imbalances, Young argues, are an essential component of a theory of justice, and are important to a complete understanding of justice.

Young worries that "a distributive understanding misses the way in which the powerful enact and reproduce their power."[15] Young's explanation of the weaknesses of the paradigm is astute, and points to a widely shared objection to many twentieth-century contributions to liberal political theory.[16] Power cannot be redistributed. To suggest otherwise is to assume an overly individualistic conception of human identity, and to omit the relationships that comprise human society.

Young also worries about identifying justice exclusively with ideal distribution: "What marks the distributive paradigm is a tendency to conceive of social justice and distribution as coextensive concepts."[17] Young accepts that distribution is part of the story of justice. Her objection to the "distributive paradigm" does not deny that some relationship exists between political injustice and unfair or unequal distributions of objects. She even suggests that the relationship may be one of entailment or causation: if oppression and domination come to be properly understood as the primary social components of injustice, distribution can be more appropriately formulated as derivative of power relations.[18]

Her objection is to the suggestion that abstract distributive concepts should be understood as the totality of a theory of justice, or, moreover, that the metaphor of distribution should be extended to include intangibles such as power relations or rights. Young's elucidation of the distributive paradigm of justice entails a multidimensional understanding of injustice. Justice may have distributional origins, or it may have distributional effects, but justice involves power, social relations and capacities in addition to distributions.

The first worry leads to the second. Where distributional analysis is broadened to explain imbalances of power or rights, Young explains that "the logic of distribution treats nonmaterial goods as identifiable things or bundles distributed in a static pattern among identifiable, separate individuals."[19] Where the distributive paradigm is extended in this way to include distributions of rights or powers, the metaphor of distribution conceals problematic assumptions. Of deep concern is the worry that the distributive paradigm obscures the role played by institutions in perpetuating injustice, and moreover that it tacitly accepts the status quo in social practices as though this were a set of background conditions unrelated to injustice. Young explains that the institutional context she is referring to "includes any structures or practices, the rules and norms that guide them, and the language and symbols that mediate social interactions within them, in institutions of state, family and civil society, as well as the workplace."[20]

Young wants to refocus theoretical analysis of injustice on the social structures and processes that produce distributions, rather than on the distributions themselves.[21] Young worries that the distributive paradigm's exclusive focus on patterns and distributions of material bundles, "ignores and tends to obscure the institutional context within which those distributions take place."[22] In particular, the distributive paradigm obscures the injustice of institutional power structures that ultimately undergird unjust distributions. Injustice has distributional outcomes, and distributional implications, but ought not to be defined solely in terms of distributions—material or otherwise.

Instead, Young argues that we should broaden our understanding of justice, and see oppression and domination as presenting a more perspicuous paradigm of injustice.[23] Distributive language should be explicitly limited to discussions of material goods, and "justice" should be understood to have a wider scope including distributive issues, but also including "procedural issues of participation in deliberation and decision-making."[24] She defines oppression as a family resemblance concept, which she calls the "five faces of oppression": exploitation, marginalization, powerlessness, cultural imperialism and violence.[25] Oppression is a set of institutional processes and systematic limitations on self-development of individuals marked as members of a socially subordinate group.[26] "Domination" involves a set of institutional conditions that prevent individuals from controlling their own actions, or from controlling the conditions of their own actions.[27] Since unjust distributions of goods can be seen as the outcome of social processes, we can use an examination of oppression and domination to better understand *how* and *why* unjust distributions may be the outcome of injustice, but without focusing our attention on distributions of goods per se.

An upshot of Young's analysis is that mere redistribution is not sufficient to remedy injustice. Bad distributions are only part of the problem, so fixing the distributions of goods alone does not suffice to make the society just. Remedies to injustice would be remiss if they ignored distributional analysis, but uncritical redistribution may reinforce and entrench existing power structures, and may amount to blind acceptance of the atomistic assumptions underlying the distributive paradigm. Mere transfer of goods alone will not ultimately impact the power imbalances that define injustice, and cause unfair and unjust distributions, so redistribution alone would not remedy the ultimate injustice understood as oppression and domination. That is, although it may aim to remedy injustice, mere redistribution leaves the underlying injustice untouched.

II. FOOD DESERTS

In this context I argue that food deserts are a problem of injustice that is most effectively defined and most readily visible as a distributive problem. Food deserts are neighborhoods in which nutritious food is scarce, of low quality, expensive, or simply unavailable.[28] These are contrasted with food oases or neighborhoods where supermarkets are present, and nutritious food is plentiful or accessible. Embedded in both concepts is the idea that access to nutritious foods varies between populations within a country, state, or city. In other words, *distribution* of foods—especially accounting for correlation between population data and type of store or quality of food—defines food deserts. Food deserts are therefore, by definition, a distributive issue. In that sense, they serve as an interesting and useful example for illustrating Young's theory.

In part, food deserts are a useful example because food deserts exemplify both unfair distribution and correlation with oppression and domination. That is, food deserts have all of the social components of Young's broadened definition of justice. The existence of food deserts correlates with low-income populations, racialized populations, disabled populations and otherwise marginalized populations.[29] More generally, variety and quality of available food have been found to correlate with affluence, and lack of availability to correlate with poverty.[30] Speculation abounds as to why this might be the case, but a view (empirically verifiable but mostly unsupported) by retailers that supermarkets face both profitability and security obstacles to locating in low-income urban environments may play a role in perpetuating existing food deserts once they emerge.[31] Food deserts are therefore yet another form that marginalization within affluent societies may take. Hence, relying on Young's five faces of oppression, food deserts may be understood as an indicator of oppression in industrialized societies.

Food justice and the food justice movement combine concern for localism and sustainability in food production with concern for inequality and social justice.[32] Sbicca explains that food justice "pursues a liberatory principle focusing on the right of historically disenfranchised communities to have healthy, culturally appropriate food, which is also justly and sustainably grown."[33] In a sense, food deserts are an easily identifiable indicator of a form of social injustice. That is, they may serve as a diagnostic tool for identifying injustice. Food deserts, in the sense understood by food justice movements, are best understood as the outcomes of unjust, hidden processes of food industrialization and globalization.

Hence, I argue that the move from an examination of food deserts to questions of food justice parallels Young's critique of the distributive paradigm. Food deserts, like their abstract analogue of unequal distributions of resources under the distributive paradigm, are relatively easy to identify by empirical means. They are also a useful indicator of injustice. They are not, however, the totality of the injustice in question. The food justice movement worries about the hidden institutions of food production and distribution, about industrialization and about globalization of both processes. Applying Young's lesson regarding the distributive paradigm to the example of food deserts, I argue that unequal distribution of food frequently coincides with injustice, but the totality of injustice goes much deeper. The underlying injustice remains one of oppression and domination, and food deserts can be seen as occurring downstream from the principle injustice.

Given the health consequences of food insecurity, food deserts can be also seen as a *cause* of further unjust social processes and oppressions. That is, they are a distributive problem that is the outcome of institutions of industrialized food processes and globalization of food production, but they are also causally related to perpetuating the injustice because they may be a reinforcing component of oppression and domination. Food deserts are associated with lower quality diets and higher rates of obesity compared with food oases.[34] They correlate with lower test scores, violence and higher rates of unemployment. In other words, they correlate with loss of opportunity, violence, and marginalization. Hence, they correlate with further oppression and domination on Young's understanding of these concepts.

Accordingly, I argue that food deserts, food security and access to nutritious food should be examined as part of a comprehensive analysis of injustice, and ought to be considered in their distributive formulation. Food deserts in their distributive formulation serve two useful purposes with respect to injustice: a diagnostic purpose, and a prescriptive or normative purpose, and it is to these two purposes that I turn next.

The observation that motivates this chapter is that, at its heart, the very idea that access to nutritious food varies between populations is a distributive one. But this distributive observation is only the beginning of the analysis of the injustice of food deserts. The example of food deserts prompts a Young-inspired acceptance of distributive analysis, even if such analysis comes alongside a rejection of strict distributive definitions of injustice. Like food deserts, a variety of justice problems are identifiable most readily in their distributive formulation, and an examination of the correlation between oppression and distribution can therefore be a useful demonstration of the urgency of certain justice claims. The discovery of unequal distributions can therefore serve the diagnostic purpose of identifying injustice.

And in many cases, as in the example of food deserts, unjust distribution is also not an endpoint to the injustice. Power imbalances are both causes of and caused by food injustice. Hence, distributional analysis is not just an effective way to see the injustice of food deserts: redistribution will also be a necessary part of the solution to these forms of injustice. A response to injustice will be inadequate if it does not at least remedy the unfairness of the distribution. When we accept Young's insight that distributions are the outcome of social processes, I conclude that we can still gain some insights by applying distributive analysis alongside social justice analysis.

In the case of food injustice, the distributive analysis both serves as an effective way to discover the presence of injustice, and suggests a partial solution to the problem. Distribution is relatively easy to measure, and maldistribution is often relatively straightforward to highlight. Better food distribution, and erasing of food deserts, is predicted to have an impact on childhood and cognitive development and adult health.[35] And these in turn have been argued to impact power relations.[36] Hence, injustice understood in terms of power relations will be both a cause and a net effect of food deserts.

III. CONCLUSION

We need not define injustice as a solely distributive problem to accept the utility of distributive analysis for diagnosing real-world examples of injustice. And although redistribution may not be sufficient to remedy injustice, redistribution is nonetheless a necessary part of the solution to some forms of injustice. This conclusion is perfectly compatible with Young's objection to conflating the distributive paradigm of justice with the totality of justice. Moreover, redistribution of certain goods, to certain populations, may have a positive impact on domination and oppression. Hence, I argue that some forms of injustice are most visible through the lens of distribution, and furthermore may be partially remediable through redistribution, and that distributive analysis is therefore worth pursuing for social justice reasons. Critical analysis of both causes and effects of unequal distributions will be required in order to more fully understand the power structures that underlie the oppression and domination. But the account of justice we are left with will be pluralist: injustice involves both unfair distributions and entrenched, hierarchical power structures. Justice will require work in response to both facets of the problem.

Ultimately, Young's objection to the distributive paradigm is an objection to identifying unfair or unequal distributions with injustice. Under that formulation, I wholeheartedly agree with Young's rejection of any political theory suggesting that justice and injustice simply or straightforwardly consist entirely of distributions. However, I would argue, and I hope I have shown, that distribution nonetheless contributes to justice broadly understood as oppression and domination, and in that sense, that distributive analysis and the distributive formulation of injustice have an important role to play in describing and diagnosing injustice. Once the diagnostic task of identifying injustice is complete, redistribution may also play a role in remedying injustice. Hence, distributive analysis remains both descriptively valid—in diagnosing injustice—and prescriptively required as part of a response to injustice.

NOTES

1. For the invitation to participate in this volume, for her careful and thoughtful comments, and for a very helpful APA commentary, I would like to thank Jill Dieterle. For thoughtful responses and questions, I'd like to thank an anonymous reviewer. For helpful responses, questions, and discussion, I would like to thank audiences at NASSP, CSWIP, APA Pacific, and CRÉ at l'Université de Montréal, and in particular Chris Lowry, Mark Navin, Christine Tappolet, Sally Scholz, Tracy Isaacs, Barrett Emerick, and Susan Dieleman.

2. Iris Marion Young, *Justice and the Politics of Difference* (Princeton: Princeton University Press, 1990), 5.

3. Environmental injustice—broadly the correlation between, on the one hand, social marginalization along racial, gender, ageist, or ableist lines and, on the other hand, exposure to toxicity and environmental burdens—forms the underlying structure against which food injustice is defined. Both emerge out of a social justice analysis of social geographic data.

4. Food deserts remain a contested concept, including but not limited to their precise definition. One debate regards whether to define food deserts in terms of the nutritional content of available ingredients, or in terms of calorific content and selection. The term may, therefore, appear misleading, since neither definition implies the absolute absence of food in these geographically defined areas. The absence of supermarkets is typically compensated somewhat by the presence of smaller convenience stores, bodegas, or fast food outlets. See Deja Hendrickson, Cherry Smith, and Nicole Eikenberry, "Fruit and Vegetable Access in Four Low-Income Food Deserts Communities in Minnesota," *Agriculture and Human Values* 23 (2006): 371–83; Renee E. Walker, Christopher R. Keane, and Jessica G. Burke, "Disparities and Access to Healthy Food in the United States: A Review of Food Deserts Literature," *Health & Place* 16, no. 5 (2010): 876–84; Katarzyna Budzynska et al., "A Food Desert in Detroit: Associations with Food Shopping and Eating Behaviors, Dietary Intakes and Obesity," *Public Health Nutrition* 16, no. 12 (2013): 2115.

5. J. P. Block, R. A. Scribner, and K. B. DeSalvo, "Fast food, Race/Ethnicity, and Income," *American Journal of Preventive Medicine* 27, no. 3 (2004): 211–17.

6. Julie Beaulac, Elizabeth Kristjansson, and Steven Cummins, "A Systematic Review of Food Deserts: 1966–2007," *Preventing Chronic Disease* 6 (2009): 1–10.

7. Andrew R. Ness and John W. Powles, "Fruit and Vegetables, and Cardiovascular Disease: A Review," *International Journal of Epidemiology* 26 (1997): 1–13; K. J. Tobin, "Fast-Food Consumption and Educational Test Scores in the USA," *Child: Care, Health, and Development* 39 (2013): 118–24.

8. Young, *Justice and the Politics of Difference*, 12–18.

9. Robert Nozick, *Anarchy, State, and Utopia* (New York: Basic Books, 1974); A. John Simmons, "Historical Rights and Fair Shares," *Law and Philosophy* 14 (1995): 149–84; Nancy Fraser and Axel Honneth, *Redistribution or Recognition? A Political-Philosophical Exchange* (New York: Verso, 2004).

10. Young, *Justice and the Politics of Difference*, 3.

11. Environmental justice is a prominent example. See, for example, Adrian Martin, Shawn McGuire, and Sian Sullivan, "Global Environmental Justice and Biodiversity Conservation," *The Geographical Journal* 179 (2013): 122–31; David Schlosberg, "Reconceiving Environmental Justice: Global Movements and Political Theories," *Environmental Politics* 13 (2004): 517–40.

12. Young, *Justice and the Politics of Difference*, 19.

13 Ibid., 16.

14. Ibid., 18.

15. Ibid., 32.

16. Fraser and Honneth, *Redistribution or Recognition*, 11, 12–13.

17. Young, *Justice and the Politics of Difference*, 16.

18. Ibid., 22.

19. Ibid., 8.

20. Ibid., 22.

21. Ibid., 18.

22 Ibid., 21–22.

23. Ibid., 33.

24. Ibid., 34.

25. Ibid., 40.

26. Ibid., 37–38, 40.

27. Ibid., 38.

28. Dave Weatherspoon et al., "Price and Expenditure Elasticities for Fresh Fruits in an Urban Food Desert," *Urban Studies* 50 (2013): 88; Nathan McClintock, "From Industrial Garden to Food Desert: Demarcated Devaluation in the Flatlands of Oakland, California," in *Cultivating Food Justice: Race, Class, and Sustainability*, eds. Alison Hope Alkon and Julian Agyeman, (Cambridge: MIT Press, 2011), 152.

29. Weatherspoon et al., "Price and Expenditure," 91.

30. Christina Black et al., "Variety and Quality of Healthy Foods Differ According to Neighbourhood Deprivation," *Health & Place* 18 (2012): 1295.

31. Weatherspoon et al., "Price and Expenditure," 90; Benjamin Shepherd, "Thinking Critically About Food Security," *Security Dialogue* 43 (2012): 196.

32. Joshua Sbicca, "Growing Food Justice by Planting an Anti-Oppressive Foundation: Opportunities and Obstacles for a Budding Social Movement," *Agriculture and Human Values* 29 (2012): 456.

33. Sbicca, "Growing Food Justice," 456.

34. Weatherspoon et al., "Price and Expenditure," 89.

35. Jim Stevenson, "Dietary Influences on Cognitive Development and Behavior in Children," *Proceedings of the Nutrition Society* 65 (2006): 361–65; Walker et al., "Disparities and Access."

36. Sbicca, "Growing Food Justice"; Gerda R. Wekerle, "Food Justice Movements: Policy, Planning, and Networks," *Journal of Planning Education and Research* 23 (2004): 378–86.

Part II

FOOD SYSTEMS

Chapter 5

Food Sovereignty

Two Conceptions of Food Justice

Ian Werkheiser, Shakara Tyler and Paul B. Thompson

The term "food sovereignty" was disseminated in the late 1990s by La Via Campesina (LVC), an organization of small-scale farmers, farm workers, indigenous communities, and other self-described "peasants" who see their individual and community identities as co-constituted with their food practices, and further see food as intimately bound up in the exploitative institutions of global capitalism.[1] In 2007, LVC organized a meeting of grassroots activists at the Forum for Food Sovereignty in Sélingué, Mali. The meeting was called Nyéléni, after a legendary female Malian peasant. The highly influential Nyéléni synthesis report lays out six principles of food sovereignty, including putting control locally in a way that recognizes traditional local territories, building knowledge and skills for the communities (thereby respecting traditional cultures without forcing communities to be museum cultures in order to sustain their ways of life), and working with nature.[2] (For a full list of the principles, see the appendix.) Food sovereignty was originally much more of a term for activists than scholars or policy makers, but this is changing as scholars have begun thinking through its implications,[3] as steps have been made toward getting food sovereignty recognized in the United Nations,[4] and as several countries in South America have added clauses protecting food sovereignty into their constitutions.[5]

At the same time as food sovereignty has been gaining ground in the discourse on food justice, usage of the phrase has become ambiguous. Consistent with its original meaning, projects to protest, fight and ultimately reform political institutions that oppress the food-related interests of subsistence farmers are advanced under the banner of food sovereignty. Closely related to this usage, efforts to advance the interests of socially, economically and politically marginalized individuals might also be included, especially when their marginalization is tied to racial, class, or gender oppression and repression, including the intersectionality of these components. Here food sovereignty may function less as a substantive doctrine about the nature, quality, or approach to the production and distribution of food than as a call for reforming dominant institutions to increase participatory justice, defined as individual and community stakeholders having more meaningful opportunities to affect the food systems in which they are embedded. This notion of food sovereignty is often proposed *in tandem* with food security. Food security is, in turn, conceptualized as reliable access to healthy and culturally appropriate food. Defined thusly, food security has long been understood as a basic human right whose moral warrant rests on mainstream conceptualizations of distributive justice. Access to food might be secured through gift or grant programs or alternatively through general income supports that guarantee adequate income for the purchase of food. However, traditional approaches to food security (such as soup kitchens, food pantries and other types of direct food distribution) deny meaningful voice to the recipients and tend to foster dependency through paternalistic relations. Even if gifts of food (or opportunities for purchase) match cultural expectations of those receiving food, they are not provided with an opportunity to participate in the design or operation of institutions for achieving food security. In this context, food sovereignty would augment the notion of food security by insisting upon procedural norms.

On the other hand, food sovereignty may refer to radical democratic projects of resisting and building alternative institutions to the dominant ones in society, using food (broadly defined) as a focus. Under this conception, food sovereignty requires resistance to dominant institutions, whether they be configured in terms of charitable gifts, state-based food entitlements or some form of market access. Food sovereignty would be conceived as implying the construction of new intra- and intercommunity relationships of resistant institutions. The emphasis on food (as opposed to other goods) calls attention to the importance of food production and consumption and to unique opportunities for organization and institution building outside of and against the capitalist food system of the larger society. Since existing mechanisms for achieving food security will reflect the institutional structure of the dominant culture, this conception of food sovereignty of necessity opposes those mechanisms, even while it does not contest the moral significance of secure access to food. In the discussion that follows, we will refer to this as *radical* food sovereignty, and we will draw out a number of contrasts with *participatory* food sovereignty that does not contest the legitimacy of existing institutions.

It is worth emphasizing that both conceptions of food sovereignty agree on the following point: one of the most important ways in which communities are subjected to injustices is through the lack of participation in decisionmaking about their food systems. Since its inception the discourse of food sovereignty has asserted that participatory justice is a goal for the movement, as well as instrumentally useful for other kinds of justice.[6] However, a distinction can be made between the different ways the two conceptions of food sovereignty see and pursue the goal of participation, and we will argue that this distinction is not clearly articulated in the current literature on the movement. This distinction can be valuable analytically for scholars working to understand the tensions within the literature on food-related social movements, for activists seeking to facilitate change and coordinate plans around shared goals, and for scholars and activists seeking productive relationships around shared projects. As scholars, practitioners and activists seeking all of the above-mentioned values ourselves, we hope this analysis illuminates the critical need to continue the discourse from multiple angles.

Section I of this chapter explores the rationale and implications for looking at participatory food sovereignty as a progressive project focused on improving participatory justice. Section II reviews sources that approach food sovereignty as a radical project. Section III discusses some of the implications of these different conceptions, looking at how they have incompatible or opposing implications and how they can nonetheless work together. Ultimately, this chapter analyzes food sovereignty as a vital but often misunderstood element of the food justice movement which, in its different conceptions, makes different claims about justice.

I. PARTICIPATORY FOOD SOVEREIGNTY AS A PROGRESSIVE PROJECT

In their food justice anthology *Cultivating Food Justice: Race, Class, and Sustainability*, Alison Hope Alkon and Julian Agyeman draw a parallel between the Environmental Justice (EJ) movement and the modern food justice movement.[7] They say that where EJ has the dual components of "equal protection from environmental pollution and procedural justice," food justice is similarly concerned with two components. Food access mirrors the pollution concern in EJ and refers to "the ability to produce and consume healthy food," while food sovereignty reflects the concern for procedural justice, and refers to a community's "right to define their own food and agriculture systems."[8] Food sovereignty is concerned with issues such as the fact that "communities of color have been subject to laws and policies that have taken away their ability to own and manage land for food production."[9] This vision of participatory food sovereignty as akin to EJ's commitment to participatory justice is a common one, and one which commits food sovereignty to a particular progressive version of participatory justice. Notably, the EJ movement arose from the historic civil rights movement that politically mobilized around civil rights issues such as equal education, voting and rights to service in public spheres.[10] Therefore, the emphasis of participatory justice was heavily influenced by the civil rights rhetoric and strategizing for equal inclusivity within public institutions. The EJ approach to participatory justice generally involves working with, pressuring, and making claims on institutions that already exist within the dominant society. Protecting their community from injustices and harms brought about by unjust institutions is a key objective for advocates of EJ. Thus, even while EJ advocates participation within the existing institutions of society, advocates of EJ also understand themselves as focused on progressively reforming those institutions "from the outside."

As mentioned earlier, food sovereignty is still underexamined in academic literature, and so reviewing the literature on EJ is a useful approach. The advocacy of progressive reforms under the banner of food sovereignty can be seen as an extension of EJ and food justice. EJ argues that environmental damage and risks are often unjust because affected stakeholders did not participate meaningfully in the decisions leading to those problems.[11] Thus it calls for participatory justice where "decision makers" are affected stakeholders, both because that process itself is more just, and because it will lead to fewer distributive or recognition injustices.[12] The National Research Council calls for involving communities "early and often"[13] in risk decisions, as part of gaining their consent. For our purposes this can be understood as, at minimum, the community accepting a decision, agreeing not to oppose it, and accepting the fact of that decision (if not its moral justification) in future decision-making processes. This can make risk characterizations and decisions "more democratic, legitimate, and informative."[14] For EJ, then, community participation is largely about working with existing social institutions to protect the community from environmental harms, and working to reform those institutions so that they will be more just and beneficial in the future. Similarly, the food justice movement gained momentum during this era of political mobilization and unrest and began to address unjust or ineffective institutions systemically rooted in the ills of what is considered our "broken food system": racism, sexism and classism.

When all this is applied to food sovereignty, we see that Alkon and Agyeman's description as the "right to define their own food and agriculture systems"[15] is a procedural norm that would be exercised through meaningful participation in decisions made with dominant social institutions. One example of this is the food sovereignty law in Nicaragua.[16] A group of NGOs, activists and elected representatives came together in 2006 to draft a law pursuing food sovereignty, with clauses calling for the protection and development of traditional seed varieties, limited food imports and bans on GMO foods, and land reform including communal land titles. This pursuit of food sovereignty through legislative means is itself based on a conception

of food sovereignty as a progressive project of participatory justice, and many of the elements of the law are based on participatory justice claims. For example, the law includes clauses criticizing and resisting international trade agreements that were entered into without meaningful participation by those affected and work to undercut democratic local control of food systems.[17] In this particular case, the law was amended and debated between various political parties and other stakeholders, particularly the Food and Agriculture Organization (FAO), who issued a set of guidelines and changes which weakened the law and which were largely adopted. In the end, though many advocates of food sovereignty were dissatisfied with the outcome, there was certainly a more meaningful role for affected groups, and from a progressive standpoint, it was a step in the right direction given the initial participatory process. Moreover, the FAO has since agreed to provide a forum for debating food sovereignty as opposed to food security as a goal they ought to pursue.[18]

For this conception of participatory food sovereignty, a lack of participation by exclusion from defining the food system is the central injustice, and dominant institutions must be reformed to make them more inclusive and just, often through means familiar to people in the civil rights and other progressive movements: protests, legislative reform, court challenges, electoral politics and development aid. An example of the last strategy is reflected in Action Aid's food policy analyst Magdalena Kropiwnicka's comment on the marginal position women farmers possess in global agriculture. She says, "Women produce up to 80 percent of the food in some developing countries on a day to day basis but they own two per cent of the resources allocated to agricultural enterprise and will struggle to grow more food unless rich countries massively increase their support."[19] An analysis based on participatory food sovereignty is compatible with a justice claim on support from rich countries, but this conception of food sovereignty will also emphasize the broader criticism of the FAO ignoring women farmers in the above example—not only their needs, but their voices. Merely trying to plug the needs of women into a political space where policy efforts typically marginalize women is both unlikely to succeed, and still does not address the justice claims these women have to meaningfully influence policies affecting their lives. Instead, participatory food sovereignty might call for the FAO, as an international agency with a mandate to reduce hunger, to acknowledge and reaffirm the centrality of women farmers to global agriculture by allowing them to have a meaningful role in shaping the FAO's policies.[20] Given the emphasis on enhancing opportunities for participation *within* the framework of existing institutions, participatory food sovereignty can, in this sense, be seen as a part of the dominant society or at least as aiming toward being a part in the future (albeit a part working hard to reform the rest). As we will see, this is in contrast to the other vision of food sovereignty as standing outside of and in opposition to dominant society.

II. RADICAL FOOD SOVEREIGNTY

As noted briefly earlier, a different discourse around food sovereignty has emerged among those advocating for greater food justice, led originally by peasant and indigenous rights movements as well as the antiglobalization movement. This more radical conception of food sovereignty explicitly sees itself as set apart from and indeed in opposition to dominant social institutions and their model of food security through industrial, capitalist food systems. While secure access to food is a goal for everyone involved in food justice discourses, the means by which this is pursued under the institutions favored by dominant social interests has been critiqued by activists coming from a more radical conception of food sovereignty. In particular, the model of food security that radical food sovereignty critiques is one in which small-scale subsistence farmers either switch to cash crops or stop farming altogether and leave the countryside to work in cities, and use the money earned thereby to buy staples grown at much larger scales in a few countries in an integrated industrial food system. As the FAO 2012 report says,[21] we must improve secure access to food for the world's poor by "creating decent jobs, paying better wages, giving them access to productive assets, and distributing income in a more equitable way." This model of food security is pursued by organizations like the World Bank and IMF when they pressure countries with financial aid and other incentives to restructure their economies along these lines, as well as by organizations like FAO, USAID, etc., which give food directly to poorer countries. Giving direct food aid may sound like an unqualified good, but critics, who call the process "dumping," claim that it reinforces the food security model by pushing small-scale farmers, unable to compete with free food, off their farms and into cities, or into export crops.[22]

Food sovereignty activists resist this model of food security. In part, this is because food security has not been successful in achieving its stated aims; famines and food shortages abound in countries that follow the model laid out for them.[23] This is perhaps unsurprising. As Sen has influentially argued, famine is not a result of a lack of food, but rather due to inequalities and a lack of "entitlements," which are often a complex admixture of money, the ability to grow food directly, a social safety net, help from one's family, and so on.[24] Reducing people's "basket of entitlements" to only their purchasing power, particularly in countries suffering from inequality in the global predatory economic system, is a recipe for vulnerability. Advocates of radical food sovereignty also critique this model because it focuses simply on having food, rather than the complex social-ecological justice questions that for these thinkers surrounds questions of food. As one critic put it, "These contemporary policies aimed at food security offer no real possibility for changing the existing inequitable, social, political and economic structure and policies that peasant movements believe are the very causes of the social and environmental destruction in the countryside in both the North and South."[25] As a final critique, implementation of the food security model is detrimental to community identity, integrity and culture. Food—its production, distribution, preparation, consumption and disposal—is always co-constituted with culture and community. The roles that a local food system plays in creating and sustaining cultural forms, economic viability, or social connections cannot be replaced by the global system. As Annette Desmarais says, "This place-bound identity, that of 'people of the land,' reflects the belief that they have the right to be on the land. They have the right and obligation to produce food. They have the right to be seen as fulfilling an important function in society at large. They have the right to live in viable communities and the obligation to build community. All of the above form essential parts of their distinct identity as peasants."[26]

In its radical guise food sovereignty is proposed as an "alternative model of agriculture and rural development"[27] providing greater empowerment and control over productive resources for communities than the institutions implied by the food security approach. Radical food sovereignty is founded on a vision of maximal democracy, committed to "(i) democratic self-determination; (ii) capacity development for individuals and groups; (iii) delivery of economic, social and/or political benefits; and (iv) the construction, cultivation, proliferation and interconnection of movements and organizations that embody the first three tenets."[28] This is not a call for just participation within existing institutions, but building alternatives outside of those institutions by strengthening communities and linking them in a network of solidarity and mutual aid.[29] Interactions with the institutions that dominate local production and exchange relations in order to serve distant owners of capital might be tolerated in this model, but only for the benefit of these emerging counternetworks. Actions in support of radical food sovereignty must include protest and other forms of resistance.

An example of this conception of food sovereignty as a radical project can be found in the work of Karnataka Rajya Raitha Sangha (KRRS), the Karnataka State Farmers' Association, from the south Indian state of Karnataka. KRRS is a member of LVC and explicitly promotes food sovereignty. It does that in part by trying to build "village republics" that are "a form of social, political and economic organization based on direct democracy, economic and political autonomy, and self-reliance."[30] KRRS is actively opposed to unjust social practices, such as caste discrimination.

In order to pursue food sovereignty and democratic flourishing for their communities, they engage in actions ranging from education about farming practices to direct action and property destruction aimed at international corporations such as Cargill, Monsanto (and fields growing Monsanto GM corn) and Kentucky Fried Chicken. They have engaged in actions more in the progressive line, such as protests against organizations like the WTO, but more of their energy is focused on developing an "International sustainable peasant development center" to build up traditional farming knowledge, support communities and construct just relationships between different communities via mutual aid.

Given this conception of food sovereignty a lack of autonomy and self-direction for communities is the central injustice evident in the inability to define and implement their own food system. Institutions that overshadow local initiatives must be resisted and worked around, often through means familiar to people in radical movements of direct action: property destruction, building up communities' capacities and solidarity. This version of food sovereignty intentionally divorces itself (if it ever was married) from the dominant society, and attempts to build a very different and more just vision of the future.

III. SOME IMPLICATIONS

Despite key differences, the practical separation between the two conceptualizations of food sovereignty is incomplete. One can certainly advocate for reform of dominant institutions to increase participatory justice while also building alternatives to those institutions, and be willing to resist the dominant institutions when they do not reform. Food sovereignty advocated on behalf of indigenous groups in the United States is particularly susceptible to a mindset that appears to straddle procedural and radical modes, for example. Perhaps because the tribes function under treaties with the U.S. federal government, they struggle simultaneously with maintenance of and resistance to institutions that mediate their relationship with what they perceive to be a dominant and oppressive external society.[31] To pick one example from the literature that fits squarely into both conceptions of food sovereignty, Catherine O'Neill says

> Fish, especially salmon, are necessary for the survival of the native peoples of the Pacific Northwest, both as individuals and as a people. Fish are crucial for native peoples' sustenance, in the sense of a way to feed oneself and one's family. Fish are also crucial for subsistence, in the sense of a culture or way of life with economic, spiritual, social, and physical dimensions—a way to *be* Yakama, or to *be* Tulalip.[32]

This is very much an argument of the radical kind, and as O'Neill's paper continues, when pollution has caused fish to have toxic levels of various contaminants and U.S. government agencies respond by getting people in these communities to eat fewer fish, the actions are preserving individual health but furthering the injustice and harm against the community's identity and sovereignty.[33] O'Neill goes on to say, this time expressing more of a participatory vision of food sovereignty, that this is an example of a larger problem where organizations doing quantitative risk assessments for toxic exposures and other environmental problems for these communities fail to incorporate the lived experience of community members and the knowledge and institutions of the communities themselves by failing to incorporate them in a meaningful way into the process.[34]

Another example of an attempt to navigate the tension between participatory and radical visions of food sovereignty is the "Share the Gourd" initiative led by Guayaki, a Fair Trade Certified yerba mate company. Share the Gourd aimed to "empower" South American indigenous communities to protect the rapidly diminishing rainforest.[35] As a participatory food sovereignty project, the company helped native South American tribes who were displaced from their rainforest home in the 1970s by industrial development to revive cultural identities inherently tied to the rainforest. Tribal return to the rainforest in the 1990s was marked with destitution, and Guayaki intervened to help the tribes create market value for rainforest products, while reviving their sense of pride and self-worth. This vision became rooted in a capitalistic mechanism assuming "the more mate they sold, the more jobs they would create and the more acres of rainforest they would rescue." This market response to an unjust circumstance was further rationalized by their need to be "self-sufficient, including knowledge of agro-forestry and biodiversity."[36] Thus, this progressive project illuminated the need to reform the system to reap the greater benefit of cultural preservation. As Sarah Besky argues, fair trade may be ultimately incompatible with the creation of just relations for food producers.[37] However, consistent with a more radical understanding of food sovereignty, Guyaki's strategy may not be entirely predicated on an optimism about the market; it could also be predicated on a recognition that the survival of peoples sometimes depends upon working within global institutions while also fostering self-determination and independence. Fair trade production can be pursued in the short term for community survival while alternatives are explored. As a young Ache tribal leader says, "Thanks to the work we did with Guayaki, we could prove to the Paraguyan authorities that we are capable as an indigenous community of developing and running a model project and making use of the forest."[38] The continuation of a fair trade strategy may or may not spill over into the radical dimension of food sovereignty, but this specific example does exemplify the potential coexistence between the two kinds of food sovereignty.

The land tenure crisis and the grassroots activism surrounding it in Detroit is an example which also includes both narratives of food sovereignty, but that emphasizes the radical conception. The Detroit Community Land Trust Working Group, Community Development Advocates of Detroit (CDAD) and the People's Platform are all strategizing around the land tenure issues in the city ranging from city council imposed zoning policies, land grabs justified as urban greening "development," and city resident rental of vacant lots for food production.[39] The ultimate goal is to create a space where the city residents are able to influence the political process in determining urban land use issues that most affect them. The procedural justice component fits well within a participatory concept of food sovereignty. Yet, while Detroit residents seek to influence the political system that governs land rights in the city, they do so to achieve a radical project where collective and cooperative ownership of land rights become vehicles to community flourishing and self-determination. In this radical vision, food sovereignty is understood as food produced and consumed by those who are most affected by food insecurities on a microscale or food injustices on a macroscale. What distinguishes this example from the previous one, beyond geographic, political, social and economic facets, is the community's anticapitalistic self-development and flourishing solution to the global corporate food regime where land tenure is consolidated and privatized in the hands of a predatory wealthy minority. While increased food security can likely result from the grassroots activism of "taking back the land," this land tenure crisis opposes the typical food security model that produces paternalistic food secure environments diminishing the autonomy and self-determination of the population being served.

IV. CONCLUSION

In this chapter we have examined the discourse of food sovereignty with the aim of exposing and distinguishing two conceptualizations of its political aims. In some cases, food sovereignty works *along with* food security, serving to augment the ethical and political rationale for providing secure access to nutritionally and culturally adequate diets with a call for procedural justice. Here, food sovereignty requires greater autonomy for individuals and groups in devising and implementing the institutions for achieving food security. The thrust of food sovereignty is to call for more participation of affected parties and client groups in food system governance, especially in connection with administrative procedures for ensuring access to food. It is worth stressing the fact that such procedures themselves will vary depending on social context.

Participatory food sovereignty for smallholders in rural areas will have a different cast than participatory food sovereignty for urban populations that have depended on food banks or income supplements to obtain access to food.

But food sovereignty has also been a term of art for more radical forms of political activism in the food system. Here, to advocate food sovereignty is to oppose or resist the institutions that are increasingly prevalent in the global food system. Given the fact that most mechanisms for ensuring food security presuppose the existing configuration of property rights and exchange relations, this form of food sovereignty is much more likely to require forms of radical change in social institutions that would render many of the moral and political arguments for food security nugatory. As a platform that has emerged from grassroots political movements actively engaged in political action, this more radical type of food sovereignty is neither obvious nor unambiguous in its deeper political commitments. Left-leaning social theorists such as Harriet Friedmann[40] or Philip McMichael[41] have long argued that capitalistic institutions are incompatible with food justice. In connecting with these critiques, radical food sovereignty might align itself with Marxist or anarchist political theory, or within a different framing, such as decolonial, postcolonial and poststructuralist theories that aim to deconstruct and (re)construct societal norms and structures. Alternatively, radical food sovereignty might commit to some form of nonideal political activism that eschews comprehensive treatments of social justice in lieu of more particular characterizations of the *in*justice evident in these concrete situations. The Marxist theoretical alternative at least has generally been thought to require total rejection of the assumptions and normative commitments implied by liberal conceptions of participatory justice.

On such a philosophical interpretation of justice, the participatory and radical conceptualizations of food sovereignty must be deemed ultimately incompatible. Nonideal theories, however, are more tolerant of ambiguity, and leave open the possibility that real change in institutions can precipitate transformations in our thinking, as well.[42] On this view, the current tension between participatory and radical food sovereignty does not necessarily imply any final incompatibility.

Clearly, a more detailed development of the political commitments entailed by radical food sovereignty lies beyond the scope of the present chapter. As our discussion of food sovereignty discourse illustrates, the practical objectives of activist groups often straddle the philosophical gap between liberal reforms that aim to enhance participation and radical resistance that signals the wholesale inadequacy of existing institutions. Our attempt to expose the difference between participatory and radical conceptualizations of food sovereignty is offered in the spirit of a more reflective political self-consciousness. It is not intended to imply that gap-straddling political activism is improper or inappropriate. A relatively recent arrival on the scene of food politics and food system analysis, food sovereignty should continue to be regarded as a discourse trope with emergent properties and as-yet untapped potential for meaning construction. Seen in this vein, recognizing the contrast between participatory food sovereignty and radical food sovereignty is a further step in the realization of a new consciousness regarding the ethical dimensions of food, and hopefully toward a greater realization of that normative potential.

APPENDIX: THE SIX PRINCIPLES OF FOOD SOVEREIGNTY (NYÉLÉNI SYNTHESIS REPORT 2007, NYELENI.ORG)

AT NYÉLÉNI 2007, we deepened our collective understanding of Food Sovereignty which:

1. **Focuses on Food for People:** Food sovereignty puts the right to sufficient, healthy and culturally appropriate food for all individuals, peoples and communities, including those who are hungry, under occupation, in conflict zones and marginalised, at the centre of food, agriculture, livestock and fisheries policies; and rejects the proposition that food is just another commodity or component for international agri-business.
2. **Values Food Providers:** Food sovereignty values and supports the contributions, and respects the rights, of women and men, peasants and small scale family farmers, pastoralists, artisanal fisherfolk, forest dwellers, indigenous peoples and agricultural and fisheries workers, including migrants, who cultivate, grow, harvest and process food; and rejects those policies, actions and programmes that undervalue them, threaten their livelihoods and eliminate them.
3. **Localises Food Systems:** Food sovereignty brings food providers and consumers closer together; puts providers and consumers at the centre of decision-making on food issues; protects food providers from the dumping of food and food aid in local markets; protects consumers from poor quality and unhealthy food, inappropriate food aid and food tainted with genetically modified organisms; and resists governance structures, agreements and practices that depend on and promote unsustainable and inequitable international trade and give power to remote and unaccountable corporations.
4. **Puts Control Locally:** Food sovereignty places control over territory, land, grazing, water, seeds, livestock and fish populations on local food providers and respects their rights. They can use and share them in socially and environmentally sustainable ways which conserve diversity; it recognizes that local territories often cross geopolitical borders and ensures the

right of local communities to inhabit and use their territories; it promotes positive interaction between food providers in different regions and territories and from different sectors that helps resolve internal conflicts or conflicts with local and national authorities; and rejects the privatisation of natural resources through laws, commercial contracts and intellectual property rights regimes.

5. **Builds Knowledge and Skills:** Food sovereignty builds on the skills and local knowledge of food providers and their local organisations that conserve, develop and manage localised food production and harvesting systems, developing appropriate research systems to support this and passing on this wisdom to future generations; and rejects technologies that undermine, threaten or contaminate these, e.g. genetic engineering.
6. **Works with Nature:** Food sovereignty uses the contributions of nature in diverse, low external input agroecological production and harvesting methods that maximise the contribution of ecosystems and improve resilience and adaptation, especially in the face of climate change; it seeks to heal the planet so that the planet may heal us; and, rejects methods that harm beneficial ecosystem functions, that depend on energy intensive monocultures and livestock factories, destructive fishing practices and other industrialised production methods, which damage the environment and contribute to global warming.

NOTES

1. Nyéléni, "Declaration," *Nyéléni 2007—Forum for Food Sovereignty* (Nyeleni.org, 2007); La Via Compesina, "Our World is Not for Sale—Priority to People's Food Sovereignty, WTO out of Agriculture," (viacampesina.org, 2001); "Tlaxcala Declaration of the Via Campesina" (viacampesina.org, 1996).

2. Nyéléni, "Nyéléni Synthesis Report," *Nyéléni 2007—Forum for Food Sovereignty* (Nyeleni.org, 2007).

3. Michael Menser, "Transnational Participatory Democracy in Action: The Case of La Via Campesina," *Journal of Social Philosophy* 39 (2008): 20–41; Michael Menser, "Transnational Self-Determination, Food Sovereignty, and the State," in *Globalization and Food Sovereignty: Global and Local Change in the New Politics of Food*, eds. Peter Andree, Jeffrey Ayers, Michael Bosia, and Marie-Joesee Massicotte (Toronto: University of Toronto Press, 2010); Hannah Wittman, Annette Desmarais, and Nettie Wiebe, eds., *Food Sovereignty: Reconnecting Food, Nature, and Community* (Oakland, CA: Food First Books, 2010).

4. Priscilla Claeys, "From Food Sovereignty to Peasants' Rights: An Overview of La Via Campesina's Rights-Based Claims Over the Last 20 Years," *Food Sovereignty: A Critical Dialogue*, http://www.yale.edu/agrarianstudies/foodsovereignty/pprs/24_Claeys_2013-1.pdf

5. Wittman, Desmarais and Wiebe, *Food Sovereignty.*
6. "Nyéléni Declaration 2007"—see appendix.
7. Alison Hope Alkon and Julian Agyeman, ed., *Cultivating Food Justice: Race, Class, and Sustainability* (Cambridge, MA: MIT Press, 2011).
8. Ibid., 8.
9. Ibid., 4.
10. Dorceta E. Taylor, "The Rise of the Environmental Justice Paradigm: Injustice Framing and the Social Construction of Environmental Discourses," *American Behavioral Scientist* 43 (2000): 508–80.
11. Robert D. Bullard, *Dumping in Dixie: Race, Class, and Environmental Quality* (Boulder, CO: Westview Press, 1990); Paul Mohai, David Pellow, and J. Timmons Roberts, "Environmental Justice," *Annual Review of Environment and Resources* 34 (2009): 405–30; Kristin Shrader-Frechette, *Risk and Rationality: Philosophical Foundations for Populist Reforms* (Berkeley, CA: California University Press, 1991); Kristin Shrader-Frechette, *Environmental Justice: Creating Equality, Reclaiming Democracy* (New York: Oxford University Press, 2002); Taylor, "The Rise of the Environmental Justice Paradigm."
12. Shrader-Frechette, *Risk and Rationality*; Shrader-Frechette, *Environmental Justice*; Taylor, "The Rise of the Environmental Justice Paradigm."
13. R. E. Kasperson, "Six Propositions for Public Participation and Their Relevance for Risk Communication," *Risk Analysis* 6 (1986): 275–81.
14. Paul C. Stern and Harvey V. Fineberg, eds., *Understanding Risk: Informing Decisions in a Democratic Society* (Washington, D.C.: National Academy Press, 1996).
15. Alkon and Agyeman, *Cultivating Food Justice*, 8.
16. "Ley de Seguridad Alimentaria y Nutricional," (2009) legislacion.asamblea.gob.ni.
17. Ibid.
18. Juan Nicastro, "FAO Accepts to Debate Food Sovereignty," *Eurasia Review* (2012) ISSN 2330-717X.
19. "FAO Summit Ignored Women Farmers," *Appropriate Technology* 35 (2008): 6.
20. R. Elmhirst, "Introducing New Feminist Political Ecologies," *Geoforum* 42 (2011): 129–32; Diane Rocheleau, Barbara Thomas-Slayter, and Esther Wangari, eds., *Feminist Political Ecology: Global Issues and Local Experience* (New York: Routledge, 2013).
21. FAO, WFP, and IFAD, *The State of Food Insecurity in the World 2012: Economic Growth is Necessary but not Sufficient to Accelerate Reduction of Hunger and Malnutrition* (Rome: FAO, 2012).
22. Christopher B. Barrett and Dan Maxwell, *Food Aid After Fifty Years: Recasting its Role* (New York: Routledge, 2007); Michel Pimbert, *Towards Food Sovereignty: Reclaiming Autonomous Food Systems* (London: International Institute for Environment and Development, 2009); Vernon W. Ruttan, *Why Food Aid?* (Baltimore, MD: Johns Hopkins University Press, 1993); William D. Schanbacher, *The Politics of Food: The Global Conflict Between Food Security and Food Sovereignty* (Santa Barbara, CA: Praeger, 2010); Theodore W. Schultz, "Value of US Farm Surpluses to Underdeveloped Countries," *Journal of Farm Economics* 42 (1960): 1019–30; Paul B. Thompson, "Food Aid and the

Famine Relief Argument (Brief Return)," *The Journal of Agricultural and Environmental Ethics* 23 (2010): 209–27; Wittman, Desmarais, and Wiebe, *Food Sovereignty.*

23. Pimbert, *Towards Food Sovereignty*; Albert Sasson, "Food Security for Africa: An Urgent Global Challenge," *Agriculture and Food Security* 1 (2012).

24. Amartya Sen, *Poverty and Famines: An Essay on Entitlement and Deprivation* (New York: Oxford University Press, 1983); Amartya Sen, *Development as Freedom* (New York: Oxford University Press, 1999).

25. Wittman, Desmarais, and Wiebe, *Food Sovereignty*.

26. Annette Aurélie Desmarais, "The Power of Peasants: Reflections on the Meanings of La Via Campesina," *Journal of Rural Studies* 24 (2008): 138–49, 139.

27. Ibid.

28. Menser, "Transnational Participatory Democracy," 24.

29. Carol Gould, *Interactive Democracy: The Social Roots of Global Justice.* (Cambridge, UK: Cambridge University Press, 2014).

30. Ashlesha Khadse and Niloshree Bhattacharya, "India: A Conversation with Farmers of the KRRS," *La Via Campesina's Open Book: Celebrating 20 Years of Struggle and Hope* (Viacampesina.org, 2013).

31. For example Joni Adamson, "Medicine Food: Critical Environmental Justice Studies, Native North American Literature, and the Movement of Food Sovereignty," *Environmental Justice* 4 (2011): 213–19; James M. Grijalva, "Self-Determining Environmental Justice for Native America," *Environmental Justice* 4 (2011): 187–92; E. Hoover, "Cultural and Health Implications of Fish Advisories in a Native American Community," *Ecological Processes* 2 (2013); Catherine O'Neill, "Variable Justice: Environmental Standards, Contaminated Fish, and 'Acceptable' Risk to Native Peoples," *Stanford Environmental Law Journal* 19 (2000): 3–393; Darren J. Ranco, Catherine A. O'Neill, Jamie Donatuto, and Barbara L. Harper, "Environmental Justice, American Indians and the Cultural Dilemma: Developing Environmental Management for Tribal Health and Well-Being," *Environmental Justice* 4 (2011): 221–30; David Schlosberg and David Carruthers, "Indigenous Struggles, Environmental Justice, and Community Capabilities," *Global Environmental Politics* 10 (2010): 12–35; Dean B. Suagee, "Turtle's War Party: An Indian Allegory on Environmental Justice," *Journal of Environmental Law and Litigation* 9 (1994): 461–97.

32. O'Neill, "Variable Justice," 5.

33. Ibid., 8–9.

34. Ibid., 31–33.

35. Kat Schuett, "Guayaki Invites You to 'Share the Gourd' to Empower Indigenous Communities," *Fair World Project* (Fairworldproject.org, 2014).

36. Ibid., 21.

37. Sarah Besky, *The Darjeeling Distinction: Labor and Justice on Fair-Trade Tea Plantations in India* (Berkeley, CA: University of California Press, 2013).

38. Schuett, "Guayaki Invites You," 22.

39. "Community Land Trust," http://www.detroitpeoplesplatform.org/resources/community-land-trusts/; "What We Do," http://cdad-online.org/about/what-we-do/.

40. Harriet Friedmann, "The Political Economy of Food: The Rise and Fall of the Postwar International Food Order," *American Journal of Sociology* (1982): 248–86; Harriet Friedmann, "The Political Economy of Food: A Global Crisis," *New Left*

Review (1993): 29–57; Harriet Friedmann, "From Colonialism to Green Capitalism: Social Movements and Emergence of Food Regimes," *Research in Rural Sociology and Development* 11 (2005): 227.

41. Philip McMichael, "Tensions Between National and International Control of the World Food Order: Contours of a New Food Regime," *Sociological Perspectives* 35 (1992): 343–65; Philip McMichael, "Global Development and the Corporate Food Regime," *Research in Rural Sociology and Development* 11 (2005): 265–363.

42. Elizabeth Anderson, "What is the Point of Equality?" *Ethics* 109 (1999): 287–337; Charles W. Mills,"'Ideal Theory' as Ideology," *Hypatia* 20 (2005): 165–83.

Chapter 6

Food Sovereignty and Gender Justice

The Case of La Vía Campesina

Mark Navin

One of the most exciting organizations in contemporary food justice activism is La Vía Campesina (LVC).[1] LVC represents over 200 million farmers (primarily peasants), from over seventy countries, in a fight for local control over food economies.

> The main goal of the movement is to realize food sovereignty and stop the destructive neoliberal process. . . . Food sovereignty is the right of peoples to healthy and culturally appropriate food produced through sustainable methods and their right to define their own food and agriculture systems. . . . It puts the aspirations, needs and livelihoods of those who produce, distribute and consume food at the heart of food systems and policies rather than the demands of markets and corporations.[2]

LVC insists that food justice requires broad political changes. It requires much more than the guarantee that everyone has enough nutrients to survive. Instead, LVC embraces *food sovereignty*, which requires people to have power over the food systems they depend on.

In this chapter, I focus on what LVC says about the role of gender justice in its organization. LVC says of itself that "the movement defends women [*sic*] rights and gender equality at all levels. It struggles against all forms of violence against women."[3] In an earlier document, the "Declaration of Nyéléni," LVC wrote that "food sovereignty implies new social relations free of oppression and inequality between men and women."[4]

What should we make of these sorts of claims about gender justice? What does the goal of ending gender oppression and gender inequality have to do with ensuring that peasants exercise control over local food economies? Along these lines, Flora has worried that organizations like LVC have a tendency to place "all manner of movements for liberation from oppression, from the Zapatistas to the women's movement" under the banner of food sovereignty.[5] Here, the suggestion is that LVC may advocate gender justice just because LVC embraces other liberatory struggles, even though there may not be a direct connection between these other struggles and the struggle for food justice. Werkheiser and Noll echo this worry. They say "there is the danger of a muddled movement without clear priorities, and one that is unable to move forward until some standard of ideological purity is met."[6] Social movements need all the allies they can get; they usually avoid unnecessarily turning people away. So, we might reasonably wonder why LVC insists on advocating for gender equality and for an end to all oppression against women.

My argument proceeds as follows: First, I say something briefly about food sovereignty and LVC (section I). Then, I identify three different kinds of efforts that LVC has undertaken to promote justice for women, and I argue that each of these is connected to the core goals of food sovereignty. These efforts include promoting the full participation of women in LVC's membership and leadership (section II), valuing and protecting women's knowledge and power in traditional food economies (section III), and helping women to flourish in the new post-subsistence-farming food economies that economic globalization has created (section IV). Then, I argue that the efforts I discuss in sections II and IV may require broader social and economic efforts to promote gender equality and to end the oppression of women (Section V). However, I conclude that the efforts I discuss in section III may still seem to be in tension with these more demanding ideals of gender justice, though this tension is not as obvious as it may first seem to be.

I. FOOD SOVEREIGNTY

People who advocate for food sovereignty demand more than a human right to food. They want local communities to have power over food economies.

Various human rights documents and institutions of the post-WWII era assert the existence of a human right to food. For example, the *Universal Declaration of Human Rights* (UDHR) of 1948 states that everyone has a right to adequate food, a claim that the *International Covenant on Economic, Social and Cultural Rights* (ICESCR) echoed in 1976.[7] International aid and development organizations have tried to guarantee the human right to food, often under the banner of "food security." According to the UN's 1996 World Summit on Food, food security requires that food be available, that people have economic and physical access to food, and that people have the knowledge and other means to use food to meet their nutritional needs.[8]

Food security is a noble goal—one that we are far from reaching in a world where more than a billion people live on less than $2/day. However, since the early 1990s, groups of activists from the developing world have objected that governments and NGOs who have pursued food security have been insufficiently critical of the many harms of agricultural globalization. In particular, they worry about the ways that food security discourse has been captured by efficiency-based arguments for disrupting traditional subsistence peasant farming:

> Governments and global food industries make us believe that a new era is coming, in which big companies will produce food for everyone. . . . This implies that small-scale farming is outdated: small farmers will leave their villages and settle in cities, where they will find employment in industry or services, and they will buy their food from the local supermarket that sells food from all the continents.[9]

On this model, urbanization, industrialization and cash-crop exports will generate sufficient economic growth to finance food imports; and food imports will be needed because subsistence agriculture will largely disappear.

Some critics of food security have rallied under the banner of a broader ideal of food justice: food sovereignty. Under the leadership of LVC, advocates of food sovereignty have embraced

> the right of peoples to define their own food and agriculture; to protect and regulate domestic agricultural production and trade in order to achieve sustainable development objectives; to determine the extent to which they want to be self-reliant; to restrict the dumping of products in their markets.[10]

Food sovereignty is focused on power. In particular, the power of developing societies—and even small local communities—to exercise control over their food economies. So, for example, food sovereignty advocates resist the IMF conditionality requirements that prohibit protectionism of local agriculture or that allow global capital to flow in and out of societies with ease. Unlike food security advocates, who may be content with meeting nutritional needs, champions of food sovereignty also focus on "where the food comes from, who produces it, [and] the conditions under which it is grown."[11] Among other reasons, this is because food sovereignty advocates believe that *real* food security requires food sovereignty.[12] They believe that local control of food economies makes peasant agrarian communities less vulnerable to the potentially devastating consequences of rapid changes in the prices of imported and exported foodstuffs.[13] In this way, food sovereignty advocates echo the famous findings of Amartya Sen: hunger does not result from a lack of food, but from a lack of democratic control over food.[14] Accordingly, food sovereignty requires that decisions about food ought not to be up to "powerful corporations or geopolitically dominant governments, but up to the people who depend on the food system."[15]

LVC is an international umbrella organization for peasant farmers' groups from around the world that are working together to resist various harms of agricultural globalization. Annette Desmarais, who has written extensively about LVC, says:

> The Vía Campesina emerged in 1993 as the Uruguay Round of the General Agreements on Tariffs and Trade (GAT) was drawing to a close with the signing of . . . the World Trade Organization's (WTO) Agreement on Agriculture and the Trade Related Intellectual Property Rights (TRIPs).[16]

LVC represents peasants, farm workers and agrarian communities from dozens of countries in the developing world. It has promoted "peasant internationalism" in response to the corporate face of the late twentieth century's economic globalization.[17]

Food sovereignty has implications for gender. In many developing societies, women do a majority of the agricultural work. In particular, women are often ultimately responsible for ensuring that their families have enough to eat. Their work tending small family plots is the backbone of family nutrition. However, unrestrained agricultural globalization has disrupted these practices. In many cases, economic forces have driven women from rural subsistence farming. They now have to try to participate in export-oriented agriculture or they must abandon their villages to seek employment in urban industries. Food sovereignty is about seeking justice in response to these disruptions of agricultural globalization.

II. WOMEN'S MEMBERSHIP AND LEADERSHIP IN LVC

I mentioned at the beginning of this chapter that LVC has committed itself to promoting gender equality and to fighting against the oppression of women. The most direct way that LVC has acted on its commitments to gender justice has been to reform itself, and to ensure that women can participate as equals in both the membership *and* leadership of LVC.

From its beginning, women have fought for equal roles in LVC.[18] In 1996, the Second International Conference of LVC created a Women's Working Group (later the Women's Commission) to promote the participation and representation of women in LVC.[19] At the Rome World Food Summit later in 1996, women made up almost 40 percent of LVC's delegates. They pushed for the inclusion of various proposals in the summit's declarations. In particular, they demanded that women be granted "greater participation in policy developments in the countryside."[20] In 2000, at the Third International Conference, LVC committed itself to the ideal of 50 percent participation by women at all levels of LVC, and it decided that each geographic region would henceforth be represented by one man and one woman.[21]

On the importance of women's equal participation in LVC, Patel writes that "in order for a democratic conversation about food and agriculture policy to happen, women need to be able to participate in the discussion as freely as men."[22] I agree, and I think there are a variety of reasons why this is the case. One reason why women ought to have equal access to membership and leadership in LVC is because this organization claims to act on behalf of peasant farmers, both to promote their self-governance and to resist oppressive forms of economic globalization. Since women are among the world's peasant farmers, their liberty right to self-government requires that they have equal status in LVC.[23] A second reason for women's equal status in LVC is because LVC aims to promote the moral equality of all people (including peasants) against the objectifying forces of contemporary agricultural globalization. Women's full and equal participation in LVC (including in leadership roles) is a way to treat women as equals, and to resist the dehumanizing forces of global capitalism.[24]

III. VALUING AND PROTECTING WOMEN'S ROLES IN TRADITIONAL FOOD ECONOMIES

In many developing countries, women control (or have until recently controlled) household subsistence farming operations. They manage seeds, crop rotation, tilling and harvesting. Women are also often responsible for the preparation of traditional foods that are grown on their household subsistence plots. Women frequently take pride in this work, and they may experience their roles in subsistence agriculture as sources of power in both the family and the broader community.

Food sovereignty advocates often claim that global corporate agriculture threatens to displace women from their roles in traditional agriculture. For example, Patel writes that peasant women "can find their agroecological knowledge supplanted by the technologies of industrial agriculture."[25] LVC has committed itself to valuing and protecting the contributions that women make to traditional food cultures, and to resisting the tendency of agricultural globalization to supplant women's traditional roles. In the Nyéléni Declaration, LVC wrote,

> Our heritage as food producers is critical to the future of humanity. This is specially so in the case of women and indigenous peoples who are historical creators of knowledge about food and agriculture and are devalued.[26]

Here, the fight against the harms of agricultural globalization is also a fight to preserve the epistemic authority and power of women in traditional food production.

Various cross-cultural studies have shown that the work women do in growing, harvesting and preparing food is often central to their identities and social positions.[27] So, efforts to preserve food cultures will also be efforts to preserve practices that affirm women's identities and promote their social roles.

Vandana Shiva often focuses on the need to protect women's traditional roles in household subsistence agriculture. She argues that the destructive tendencies of corporate food systems result from broader failures to recognize and value women's traditional food knowledge and practices. In particular, Shiva thinks that traditional household subsistence farming is more sustainable than large-scale corporate agriculture.[28] In her view, this is because women have an organic relationship with nature, while the destruction of women-centered traditional subsistence agriculture heralds the dawn of inherently destructive (and necessarily patriarchal) forms of agricultural production.[29]

Efforts to value and protect women's contributions to traditional food economies seem directly relevant to the central projects of food sovereignty. One way in which women are harmed by contemporary forms of agricultural globalization is that these phenomena often lead to the elimination of women's traditional roles in growing, harvesting and preparing food for their families. Therefore, efforts to value and protect women's contributions to traditional food economies are straightforwardly efforts to *mitigate* the disruptive harms of agricultural globalization. (In contrast, in the next section I discuss LVC's efforts to help women *adapt* to the disruptive harms of agricultural globalization.)

IV. HELPING WOMEN TRANSITION TO POST–SUBSISTENCE FARMING FOOD ECONOMIES

One way to help women participate in a post–subsistence farming world is to help them grow crops for sale and export. But women in many countries are not able to compete with men in commodity agriculture. The Food and Agriculture Organization of the United Nations writes that

> women are less likely than men to own land or livestock, adopt new technologies, use credit or other financial services, or receive education or extension advice. In some cases, women do not even control the use of their own time.[30]

LVC helps women adapt to agricultural globalization by promoting land reform (including land redistribution), legal reforms that encourage women's land ownership, initiatives to distribute credit and new technologies to women farmers, and broader political and social reforms that allow women to compete on fair terms with men farmers.

Another way to help peasant women participate in the economies that agricultural globalization has created is to facilitate women's transition to urban life. The decline of small-scale subsistence agriculture has led peasants to abandon the countryside in favor of urban areas in search of employment.[31] Indeed, in recent years, the vast majority of new participants in wage-paying work have been women, because of their rapid move away from household subsistence farming.[32] However, the formal labor force has not been kind to the women of the developing world. They face wage discrimination, unsafe working conditions and sexual assault and harassment. Furthermore, many of the developing world's urban areas lack effective social services, in part due to the fact that the World Bank and IMF required cuts to these services as conditions for development loans in recent decades.[33] Also, while many peasant women are drawn to urban areas by the promise of jobs in industry, they are often unable to find employment, but are forced to live in urban slums, where they are especially vulnerable.[34] Accordingly, food sovereignty movements may aim to help women adapt to new (food) economies by instituting protections for women workers and by providing urban social services to reduce peasant women's vulnerability to violence and various forms of depravation.

Before moving on, it may be helpful to reflect on Alison Jaggar's insight that women's disproportionate victimization at the hands of economic globalization results from a combination of both global and local causes.[35] Local gender oppression makes women vulnerable to discrimination and abuse when they seek work outside the home (if they are permitted to do so at all). But the forces of economic globalization have made it necessary for women to seek employment outside the home, by making subsistence agriculture and homestead production economically unfeasible. So, while we may be tempted to think of this section's adaptation strategies primarily through the lens of "reforming sexist practices in developing societies," these practices are the target of food sovereignty activism because of the destructive impact of the forces of economic globalization. And these forces are controlled by the societies (and corporations) of the developed world. At the very least, this means that members of wealthier societies should not applaud themselves for their supposed superiority on issues of gender justice. Wealthier societies are not better for women if wealthier societies are responsible for the fact that women in poorer societies are especially vulnerable to gender-based oppression.

V. RADICAL EGALITARIANISM

The boldest part of its commitment to gender justice is LVC's embrace of the ideal of women's full social and political equality. Consider the following, which I quoted at the beginning of the paper: "Food sovereignty implies new social relations free of oppression and inequality between men and women."[36] This is more than a call to make room for women in the membership and leadership of LVC. It also seems to go far beyond mitigating the harms of globalization or adapting to those changes in ways that help women. To quote Patel, LVC appears to endorse "a radical egalitarianism," which would require dramatic changes in the distribution of economic and social power between men and women, and would require a revolutionary reimagining of the family and of gender socialization, more generally.[37]

We may reasonably wonder what full gender equality and an end to gender oppression have to do with food justice. Even though there are many good reasons to endorse these goals, they may not seem to be grounded in a commitment to food justice. Therefore, we might reasonably wonder whether LVC would do better not to commit itself to such demanding goals, since there may be strategic reasons to focus its attention more narrowly.

I think this worry may be blocked. Greater gender equality and an end to (at least some forms of) women's oppression may be necessary if LVC is going to achieve some of its core goals. In particular, in the absence of successful feminist reforms, it may be impossible to guarantee women's equal status in LVC or to protect women from new vulnerabilities created by the decline of subsistence agriculture.

First, women's equal participation in LVC may not be realizable if women are hamstrung by broader social and political inequalities. On this point, Desmarais writes that "women's daily unequal access to and control over productive, political and social resources remain significant barriers to their equal participation and representation in the Vía Campesina."[38] Desmarais adds that "the gender division of labor means that rural women have considerably less access to a most precious resource, time."[39] The Women's Commission of LVC can guarantee as much formal gender equality as it wants, but its efforts will not guarantee women's full participation in LVC. This is why LVC is committed to more radical egalitarian goals than may initially seem to be justified by their concern for food justice. On this point, Menser writes:

> Because of LVC's commitment to equality in participation, it has been forced to innovate to address a most pervasive source of social and political exclusion within its base: patriarchy and sexism. . . . Mere formal inclusion is not enough; women must have equal agency in the institutional processes.[40]

Equality of agency in the institutional process of LVC requires broader social and political equality for women. Accordingly, food sovereignty movements must target social and institutional sexism, if women are going to be able to participate in shaping the policies of LVC.[41]

Second, it may not be possible to protect women from new vulnerabilities created by the decline of subsistence agriculture in the absence of efforts to combat broader forms of gender inequality. Consider the issue of women's ownership of land. Women in almost every developing society have the legal right to own land. But in some countries, like Cambodia, there are weighty cultural norms against women's land ownership. Few women own land even though the law does not prohibit them from doing so.[42] Similar norms exist for peasant women in many parts of India.[43] On a related note, a woman's formal right to inherit land and capital will not help her to compete in agricultural commodity markets if the cultural norm is for families to bequeath property only to male heirs.

Many advocates of food sovereignty are explicit that legal reforms must be accompanied by broader feminist social reforms. For example, while Carolyn Sachs argues that food sovereignty requires "land reform and redistribution," she partners that demand with a requirement that societies "rethink and redefine heteronormative household models."[44] Members of the developed world should not be surprised by the fact that legal reforms are often insufficient to protect the economic interests of women under conditions of sexism. Even though women in the developed world have made great gains in education and employment, the unequal distribution of household labor in heterosexual marriages is a persistent problem.[45] Women in developing societies face this problem, too. As they start to enter the formal labor force, they have continued to do a disproportionate share of unpaid household labor.[46] Clearly, broader feminist struggles are necessary.

I have shown that LVC is unlikely to achieve two of its gender justice goals in the absence of a more radical gender egalitarianism. Women are unlikely to achieve equal status in LVC or be protected from vulnerabilities associated with post-subsistence agriculture unless they have far greater social and economic power.

This leaves the question of how efforts to value and protect women's roles in traditional food economies might relate to broader feminist struggles. On the surface, efforts to preserve and value food traditions may seem to be at cross purposes with efforts to end the oppression of women, since traditional food economies often involve oppressive hierarchies of power. For example, we may worry that efforts to preserve food traditions are symptoms of an unreasonable romanticism—a nostalgic longing for an undesirable world that has long ago passed by.[47] I think that this objection goes too far. LVC is not trying to bring back women-led subsistence agriculture in communities where that practice is no longer feasible. In such circumstances, LVC aims primarily at facilitating women's engagement with formal labor and commodity markets. However, in many communities, women-led subsistence agriculture is not yet "the past." In such circumstances, efforts to value and protect women's roles in subsistence agriculture are intended to avoid or mitigate significant harms. Furthermore, a commitment to mitigate some of the harmful disruptions of agricultural globalization does not entail the belief that household subsistence agriculture is the best (or even a good) way to organize food economies. Rather, it entails only that one way to promote the interests of women is to resist massive immediate disruptions to practices that currently feed millions of people. So, even if such efforts do not require a commitment to radical gender egalitarianism, they are at least consistent with such commitments.

But perhaps there is something intrinsically antifeminist about efforts to value and protect women's roles in traditional agriculture. Perhaps these efforts rely on problematic claims about women's essential natures. That is, even if we accept that women's traditional roles in food cultures may sometimes give them power in the family, as some have argued,[48] we may still worry that women's uncompensated and unrecognized domestic work can "reinscribe women's subordination in the home."[49] If emphasizing women's supposed fundamental differences from men reinforces their inequality and relative powerlessness, this is a good reason to reject the claim that it is emancipatory to emphasize women's (supposed) fundamental differences from men. So, valuing and protecting women's traditional food roles may be inconsistent with broader feminist struggles, after all.

I agree that we may want to reject the gender essentialism to which Shiva (and others) seem committed. It can be both inaccurate and unhelpful to defend peasant women by claiming that they are naturally predisposed toward food work. Not all women are the same, and not all women differ from (all) men in the same ways. Furthermore, the claim that women and men have fundamental natural differences may undercut arguments for social and political gender equality that rely on claims about the common capacities, needs and interests of men and women.

Even though we may have good reasons for rejecting gender essentialism, Shiva may still be right to advocate defending women's traditional roles against the destructive power of globalization. This is a matter of nonideal theory. If we face a choice between maintaining (an admittedly sexist) system that grants women some power over the food system, and embracing a post-subsistence economy that has no good place for women workers, we may do best to act conservatively. Consider that you should usually try to stop someone from burning down your house, even if it has a leaky roof. But your efforts to save your house should not be taken as evidence that you love your house as it is. Perhaps you'd like to move, but you'd rather stay in your imperfect house than be on the street. I think we can say something similar about attempts to value and preserve women's traditional roles in food cultures. These efforts need not be antifeminist, though they sometimes may be. Indeed, they may be eminently feminist, since they may resist forces that are harmful to women.

VI. CONCLUSION

I have argued that LVC is engaged in three projects to promote gender justice. All three projects are directly connected to the core goals of food sovereignty. I have also argued that LVC's rhetoric about gender equality—and ending the oppression of women—may be justified in terms of some of its gender-justice projects. In particular, LVC will succeed in its efforts to include women in its own organization and to facilitate women's involvement in post–subsistence farming economies only if broader feminist reforms succeed.

NOTES

1. I am grateful for helpful feedback from Ami Harbin, Phyllis Rooney, and Jill Dieterle.

2. La Via Campesina, "The International Peasant's Voice," *La Via Campesina International Peasant's Movement*, February 9, 2011, http://viacampesina.org/en/index.php/organisation-mainmenu-44.

3. Ibid.

4. La Via Campesina, "Declaration of Nyéléni," *La Via Campesina International Peasant's Movement*, February 27, 2007, http://viacampesina.org/en/index.php/main-issues-mainmenu-27/food-sovereignty-and-trade-mainmenu-38/262-declaration-of-nyi.

5. C. Flora, "Book Review: Schanbacher, William D.: The Politics of Food: The Global Conflict between Food Security and Food Sovereignty," *Journal of Agricultural and Environmental Ethics* 24 (2010): 545.

6. Ian Werkheiser and Samantha Noll, "From Food Justice to a Tool of the Status Quo: Three Sub-Movements within Local Food," *Journal of Agricultural and Environmental Ethics* 27 (2014): 209.

7. Article 25.1 of UDHR states that "everyone has the right to a standard of living adequate for the health and well-being of himself and of his family, including

food," while Article 11.1 of the ICESCR states that "the States Parties to the present Covenant recognize the right of everyone to an adequate standard of living for himself and his family, including adequate food."

8. World Health Organization, "Food Security," January 21, 2015, http://www.who.int/trade/glossary/story028/en/.

9. Michel Pimbert, "Women and Food Sovereignty," *LEISA Magazine* 25 (2009): 6; see also Shelley Feldman and Stephen Biggs, "International Shifts in Agricultural Debates and Practice: An Historical View of Analyses of Global Agriculture," in *Integrating Agriculture, Conservation and Ecotourism: Societal Influences*, ed. W. Bruce Campbell and Silvia López Ortíz, Issues in Agroecology—Present Status and Future Prospectus 2 (Springer Netherlands, 2012), 107–61.

10. La Vía Campesina, *Tlaxcala Declaration of La Vía Campesina* (Tlaxcala, Mexico, 1996), http://viacampesina.org/en/index.php/our-conferences-mainmenu-28/2-tlaxcala-1996-mainmenu-48/425-ii-international-conference-of-the-via-campesina-tlaxcala-mexico-april-18-21.

11. Pimbert, "Women and Food Sovereignty," 8.

12. Annette Aurélie Desmarais, "The Via Campesina: Peasant Women at the Frontiers of Food Sovereignty," *Canadian Woman Studies* 23 (2003): 141.

13. Michel Pimbert, *Towards Food Sovereignty: Reclaiming Autonomous Food Systems* (London: IIED, 2009).

14. Amartya Sen, *Poverty and Famines* (Oxford: Clarendon Press, 1981).

15. Raj Patel, "Food Sovereignty: Power, Gender, and the Right to Food," *PLoS Medicine* 9 (June 26, 2012): 2, doi:10.1371/journal.pmed.1001223.

16. Desmarais, "The Via Campesina," 140.

17. Walden F. Bello, *The Food Wars* (London: Verso, 2009).

18. Annette Aurélie Desmarais, "The Power of Peasants: Reflections on the Meanings of La Vía Campesina," *Journal of Rural Studies* 24 (2008): 138–49.

19. Desmarais, "The Via Campesina," 142.

20. Ibid., 143.

21. Ibid., 144; Annette Aurélie Desmarais, *La Vía Campesina: Globalization and the Power of Peasants* (Fernwood Publishing, 2007).

22. Patel, "Food Sovereignty," 2.

23. For a defense of democracy that is grounded in the value of liberty, see, for example, Carol C. Gould, *Rethinking Democracy: Freedom and Social Cooperation in Politics, Economy, and Society* (Cambridge: Cambridge University Press, 1988).

24. For a defense of democracy that is grounded in the value of equality, see, for example, Jeremy Waldron, *Law and Disagreement* (Oxford: Oxford University Press, 1999).

25. Patel, "Food Sovereignty," 3.

26. La Vía Campesina, "Declaration of Nyéléni."

27. See, for example Carole Counihan, *Around the Tuscan Table: Food, Family, and Gender in Twentieth-Century Florence* (New York: Routledge, 2004); Theresa W. Devasahayam, "Power and Pleasure around the Stove: The Construction of Gendered Identity in Middle-Class South Indian Hindu Households in Urban Malaysia," in *Women's Studies International Forum*, vol. 28 (Elsevier, 2005), 1–20; Josephine A. Beoku-Betts, "We Got Our Way of Cooking Things: Women, Food, and Preservation of Cultural Identity among the

Gullah," *Gender & Society* 9 (1995): 535–55; Patricia Allen and Carolyn Sachs, "Women and Food Chains: The Gendered Politics of Food," in *Taking Food Public: Redefining Foodways in a Changing World*, eds. Psyche Williams Forcan Carol Courinan (New York: Routledge, 2012), 23.

28. Vandana Shiva, *Staying Alive: Women, Ecology and Development* (Zed Books, 1988).

29. Maria Mies and Vandana Shiva, *Ecofeminism*, 2nd ed. (Zed Books, 2014).

30. Food and Agriculture Organization of the United Nations, "Gender," 2015, http://www.fao.org/gender/gender-home/en/?no_cache=1.

31. Bello, *The Food Wars*.

32. Arabella Fraser, "Harnessing Agriculture for Development," *Oxfam Policy and Practice: Agriculture, Food and Land* 9 (2009): 56–130.

33. J. E. Stiglitz, *Making Globalisation Work* (New York: Norton, 2006).

34. Mike Davis, *Planet of Slums*, rep. ed. (New York: Verso, 2007); Nicole Hassoun, *Globalization and Global Justice: Shrinking Distance, Expanding Obligations* (Cambridge: Cambridge University Press, 2012).

35. Alison M. Jaggar, "'Saving Amina': Global Justice for Women and Intercultural Dialogue," *Ethics & International Affairs* 19 (2005): 55–75.

36. La Vía Campesina, "Declaration of Nyéléni."

37. Raj Patel, "What Does Food Sovereignty Look Like?" *The Journal of Peasant Studies* 36 (2009): 270.

38. "The Via Campesina," 144.

39. Desmarais, *La Vía Campesina*, 178.

40. Michael Menser, "Transnational Participatory Democracy in Action: The Case of La Via Campesina," *Journal of Social Philosophy* 39 (2008): 20–41. According to the Food and Agriculture Organization of the United Nations, women make up 43 percent of agricultural workers in developing societies, Food and Agriculture Organization of the United Nations, "Gender."

41. Patel concurs:

> To make the right to shape food policy meaningful is to require that everyone be able substantively to engage with those policies. But the prerequisites for this are a society in which the equality-distorting effects of sexism, patriarchy, racism, and class power have been eradicated. Activities that instantiate this kind of radical "moral universalism" are the necessary precursor to the formal "cosmopolitan federalism" that the language of rights summons. And it is by these activities that we shall know food sovereignty. (Patel, "What Does Food Sovereignty Look Like?" 270)

42. Fraser, "Harnessing Agriculture for Development."

43. Esther Vivas, "Without Women There Is No Food Sovereignty," *International Viewpoint*, February 2, 2012, http://www.internationalviewpoint.org/spip.php?page=imprimir_articulo&id_article=2473.

44. Carolyn Sachs, "Feminist Food Sovereignty: Crafting a New Vision" (Food Sovereignty: A Critical Dialogue, Yale University, 2013), 8–9, http://www.yale.edu/agrarianstudies/foodsovereignty/pprs/58_Sachs_2013.pdf.

45. Arlie Hochschild and Anne Machung, *The Second Shift: Working Parents and the Revolution at Home* (New York: Viking Press, 1989); Janeen Baxter, Belinda

Hewitt, and Michele Haynes, "Life Course Transitions and Housework: Marriage, Parenthood, and Time on Housework," *Journal of Marriage and Family* 70 (2008): 259–72.

46. Vivas, "Without Women There Is No Food Sovereignty"; Shelley Feldman, "Rethinking Development, Sustainability, and Gender Relations," *Cornell Journal of Law & Public Policy* 22 (2012): 649.

47. See, for example Paul Collier, "The Politics of Hunger: How Illusion and Greed Fan the Food Crisis," *Foreign Affairs* 87 (2008): 67–79.

48. Kurt Lewin, "Forces behind Food Habits and Methods of Change," in *The Problem of Changing Food Habits*, Bulletins of the National Research Council 108 (Washington, D.C.: National Research Council, National Academy of Science, 1943), http://www.nap.edu/openbook.php?record_id=9566&page=35.

49. Allen and Sachs, "Women and Food Chains," 3.

Chapter 7

Food Policy, Mexican Migration and Collective Responsibility

Steve Tammelleo

This chapter examines the effect of the North American Free Trade Agreement (NAFTA) on Mexican migration. Because NAFTA removed trade barriers while allowing for the continuation of agricultural subsidies, it resulted in significant agricultural job losses in Mexico. In the same year that NAFTA was enacted, the Clinton administration introduced a new border policy, Operation Gatekeeper, which made it more difficult for undocumented workers to cross the U.S.–Mexico border. Furthermore, U.S. government and industry played central roles in bringing Mexican labor to the United States for more than 100 years. As a result of NAFTA, Mexican migration increased dramatically; yet, because of Operation Gatekeeper, the number of migrants dying while attempting to cross the border also increased. Using Larry May's account of collective responsibility, I argue that the members of the U.S. Congress are collectively responsible for failing to prevent the deaths that have resulted from the combined effects of trade and border policies. In addition, I suggest policy changes that the U.S. Congress should enact to reduce Mexican migration and minimize the deaths of migrants along the border.

I. NAFTA AND MEXICAN MIGRATION

By the late 1980s, within the field of developmental economics a strong consensus had emerged that free market reforms, including the establishment of free trade and the development of a strong export sector, was the best model of economic growth for developing countries. As economist Charles Wheelan explains, free trade can be the foundation for strong economic growth. Trade leads to increases in specialization, specialization results in increases in productivity, and productivity enriches society. With trade, each nation can focus on its strengths. Trade enables the world market to utilize scarce resources in the most efficient way.[1]

On January 1, 1994, NAFTA created a free trade zone that includes Canada, the United States and Mexico. Twenty years after the passage of NAFTA, it is clear that the overall effect of NAFTA has been beneficial for Mexico's economy. As Ernesto Zedillo Ponce de Leon, president of Mexico from 1994 to 2000, points out, in the twenty years of NAFTA, annual trade between Mexico, the United States and Canada has increased 400 percent and interregional development has increased nearly 500 percent.[2] In addition, in the Mexican manufacturing sector, NAFTA resulted in the creation of 1.3 million jobs.[3] After a fairly quick recovery from a peso devaluation in 1995, from 1996 to 1999, Mexico's GDP grew at an average annual rate of 5.2 percent, and, in 2000, the rate of growth hit 6.9 percent.[4] However, the restructuring of the economy and the introduction of foreign competition that occurred with the arrival of NAFTA had devastating consequences for rural farmers in Mexico, particularly those involved in corn production.

The religious and cultural significance of corn in Mexico is tremendous. According to the Popul Vul, the Mayan creation myth, after earlier attempts to make human beings out of earth and wood had failed, the gods used corn to create human beings. For many indigenous people who live in rural areas of Mexico, corn continues to play a central role in religious and cultural life. In a 2007 interview with political scientist Judith Adler-Hellman, Don Beto, a metalworker from the village of San Jeronimo in the state of Puebla, describes the significance of planting corn:

> For me it's absolutely essential. It's the most important thing I do. . . . I am a man, so I plant corn. That's what my father did, that's what my grandfather did, and if my sons now prefer to migrate al norte than to do this work, then I can understand. Times have changed. But I haven't changed, and I don't feel right if I'm not planting and harvesting.[5]

In addition to its importance to religious and cultural life, corn continues to play a central economic role in Mexico. In 1990, 30 percent of the arable land in Mexico was used to grow corn; the value of the corn harvest was one-third of the total value of all agricultural products in Mexico.[6] Of 4 million farmers in Mexico, 2.7 million produce corn. The great majority of corn farmers are extremely poor; more than 90 percent of corn farmers grow corn on plots of land smaller than five hectares.[7]

The differences between agricultural production in the United States and Mexico are dramatic. In 2002, in Mexico, there were only 20 tractors for every 1,000 agricultural workers, while in the United States, there were 1,484 tractors for every 1,000 agricultural workers. Due to poor infrastructure, Mexican producers have to pay more for electricity, diesel fuel, transportation, and warehousing than farmers in the United States. Although Mexico has substantial subsidies that represented 22 percent of the value of agricultural production in 2001, farm subsidies in the United States were even more generous, representing 36 percent of the value of agricultural production in the same year. As a result, the average per annum productivity of an agricultural worker in the United States was $39,001, while in Mexico it was merely $2,164. Between 1999 and 2001, Mexican farmers produced an average of 2.5 tons of corn per hectare and U.S. farmers produced an average of 8.55 tons per hectare.[8] As for total annual production levels, farmers in the United States produce fourteen times more corn than farmers in Mexico.[9]

In recognition of its unique economic and cultural importance to Mexico, NAFTA established fairly tough barriers to U.S. corn exports in the first year of the trade agreement that would gradually be reduced over a fifteen-year period. According to NAFTA rules, in 1994, the United States would be allowed to export 2.5 million tons of corn tariff free to Mexico, but U.S. corn exporters would be required to pay penalties of $197 per ton for exports that exceeded this limit. In each year that followed, NAFTA increased the total quota of tariff-free corn exports by 3 percent and in 2008 trade restrictions on corn exports were eliminated.[10]

The gradual reduction of trade barriers for corn was designed to give poor Mexican corn farmers time to adjust to U.S. competition. However, when NAFTA went into effect, Mexican officials failed to enforce the tariffs outlined by NAFTA. In fact, Mexico's Secretary of the Economy offered statistics on agricultural imports that were much different than the statistics reported by the U.S. Department of Agriculture. For example, for the year 2003, the U.S. Department of Agriculture reported that 6.1 million tons of corn had been exported to Mexico, but Mexico's Secretary of the Economy reported that only 3.8 million tons of corn had been imported.[11] The regular disparities between American and Mexican statistics on agricultural imports/exports suggests that Mexican custom offices lacked the resources to adequately measure the quantity of agricultural imports. Thus, Mexican officials did not enforce the tariffs on corn imports. In 1994, U.S. farmers exported 3 million tons of corn to Mexico, slightly more than their quota of 2.5 million tons. But in 1995 and 1996, U.S. corn exports to Mexico totaled 5.9 million tons and 6.3 million tons. In the period from 1997 to 2005, with the exception of 1997, U.S. corn exports exceeded 5 million tons every year.[12] So rather than a gradual transition to competition with American farmers with the introduction of NAFTA, Mexican farmers were overwhelmed by an invasion of cheap subsidized U.S. corn.

The damage caused by NAFTA was magnified by the fact that the Mexican government did nothing to inform small farmers of the dramatic changes that would result from the trade agreement. In the years proceeding NAFTA, the Mexican government shared information about NAFTA with the country's largest agricultural companies, yet information about NAFTA was not made available to smaller agricultural producers. In its typical autocratic fashion, the Institutional Revolutionary Party (PRI), which had ruled Mexico for more than sixty years, did not value transparency, openness, or accountability. Poor farmers with small farms were left in the dark. As a result, the great majority of the 2.7 million corn farmers in Mexico did nothing to prepare for the radical changes that NAFTA would introduce.

In 1994, the average Mexican corn farmer had less than five hectares of land. He did not have a tractor. His access to agriculture research and technical expertise was extremely limited. The road network connecting his farm to major markets was inadequate. He had to pay high prices for diesel fuel and electricity. His local, state and federal governments were all corrupt, offering him merely a mirage of democracy. He probably did not graduate from the eighth grade. He may or may not have had electricity or running water in his home. The average Mexican corn farmer could not compete.[13]

In a desperate attempt to hold on, small farmers put more land into corn production; as a result, Mexico's total corn crop increased from 10.5 million tons in 1988 to 18.3 million tons in 1999.[14] In some cases, by providing technical expertise and hybrid seeds, multinational agribusiness companies helped small farmers to improve their crop yields. A group of small corn producers working with Monsanto managed to increase their crop yields from 3.7 tons per hectare to 5.7 tons per hectare.[15] Despite these efforts, NAFTA had a devastating effect upon agricultural workers in Mexico. According to the Mexican Department of Labor and Social Security, in 1993, 8,842,774 Mexican workers were employed in agricultural activities, but by 2003, that figure had dropped to 6,813,644. Thus, in the first ten years of NAFTA, more than 2 million jobs were lost in Mexico's agricultural sector.[16] Some of these workers moved from rural areas to Mexico's cities, others crossed the border to seek work in the United States. According to historian Ronald Takaki, between 1994 and 2007 more than 6 million Mexicans migrated to the United States.[17] Yet, at precisely the same time that NAFTA produced an upswing in Mexican migration, the U.S. Border Patrol introduced new policies that made crossing the border much more difficult, expensive, dangerous and in many cases deadly.

II. OPERATION GATEKEEPER

Given their centrality to politics today, it is hard to remember that in the 1970s and 1980s, Mexican immigration and control of the border were unimportant issues in both regional and national politics. The national platform of the Republican Party did not even mention immigration policy or border policy until 1980. The Democratic Party didn't discuss immigration or border policies in its national platform until 1996.[18] Up until the late 1970s, there were fewer than 2,000 Border Patrol agents nationwide.[19]

In the 1980s, due to a severe economic downturn in Mexico, Mexican migration to the United States dramatically increased. Roughly 400,000 immigrants (both documented and undocumented) arrived in the state of California annually in the late 1980s and early 1990s.[20] As a result of migration, the ethnic composition of California radically shifted. In 1960, California was one of the U.S. states with the highest percentage of white residents, but by the 2000 census, non-Hispanic whites in California were only 40 percent of the population.[21] As a corollary of the state's rapidly changing demographics, California saw an increase in nativist and anti-immigrant attitudes. In 1986, 73 percent of California voters supported Proposition 63, which made English the official language of California.[22] And in 1994, 59 percent of the California electorate voted in favor of Proposition 187, which restricted undocumented immigrants' access to education, health care and other social services. This context of rising immigration and increasing nativist sentiment provided the necessary preconditions for radical changes in U.S. border policy.

Prior to the early 1990s, the main points of entry for undocumented workers were Tijuana/San Diego and Juarez/El Paso. The main strategy of the Border Patrol was to make as many arrests as possible; in this period, the Border Patrol used the increasing number of arrests to demonstrate their effectiveness and to justify increases to their annual budgets. A radical change in strategy began in September 1993, when Silvestre Reyes, the chief of Border Patrol in El Paso, introduced Operation Blockade. Reyes's strategy was to concentrate a large number of Border Patrol agents on the banks of the Rio Grande. As a result, migrant traffic was redirected and the number of arrests in the El Paso Sector plummeted. The citizens of El Paso loved Operation Blockade and the national media praised the success of Reyes's innovative approach.[23]

In 1994, the Clinton administration decided to replicate the strategy of Operation Blockade in California. Under the title Operation Gatekeeper, this new strategy of creating deterrence by concentrating large numbers of Border Patrol agents at a key point of entry was introduced to the Tijuana/San Diego border in October of 1994. In the twelve months prior to the introduction of Operation Gatekeeper, the Border Patrol arrested 450,000 migrants in the San Diego sector, a figure that represented almost 50 percent of the total number of Border Patrol arrests nationwide.[24] Yet, within months the arrest rate in the San Diego sector of the border plummeted. The massive increase in the number of border patrol agents, the use of stadium lighting and the construction of a physical barrier along the border made crossing at San Diego much more difficult; over time immigration traffic shifted to the deserts of Arizona, New Mexico and Texas. Operation Gatekeeper was hailed as a success by politicians of both parties, yet, it was not without its critics. An ACLU petition to the Organization of American States argued:

> Operation Gatekeeper has steered the flow of immigrants into the harshest, most unforgiving and most dangerous terrain on the California-Mexico border. The United States has purposefully done this knowing that the policy would dramatically increase the number of border crossers who die and without taking adequate steps to prevent those deaths.[25]

With NAFTA destroying jobs within Mexico and the two main crossing points effectively blocked, the flow of immigration was shifted to the deserts of Arizona, New Mexico and Texas. And so, the dying season began. According to anthropologist Ruth Gomberg-Munoz the introduction of Gatekeeper resulted in a 500 percent increase in the number of migrant deaths.[26] In his book on Operation Gatekeeper, geographer Joseph Nevins claims that from 1995 to 2009, more than 5,000 bodies were recovered in the U.S.–Mexico borderlands, roughly one recovered body per day for fifteen years.[27]

The years following Gatekeeper saw a dramatic increase in the size of the Border Patrol. From 1993 to 2013, the number of Border Patrol agents increased from 3,389 to more than 21,000.[28] As a result of greater control over the border, the total number of Border Patrol apprehensions has been rapidly declining, from roughly 1,650,000 in 2000 to less than 400,000 in 2012.[29] However, the number of bodies recovered in the desert each year has stayed at the same level; from 2000 to 2012, between 300 and 500 bodies have been recovered each year. As more Border Patrol agents have been added each year, migrants have been forced away from cities and highways into increasingly remote areas of the desert. Thus, as the number of migrants crossing the border has fallen while the number of deaths has remained constant, the percentage of migrants who died during their attempt to cross the border has more than tripled in the last ten years.[30]

The stated goal of Operation Gatekeeper was to make it so difficult for migrants to enter the United States that it would act as a deterrent and persuade many migrants to stay in Mexico. In addition to those who died, increases in smuggler fees that resulted from Gatekeeper have made migrants more vulnerable to kidnapping and the abuses of modern slavery. According to journalist Terry Greene Sterline, roughly 13,000 migrants are kidnapped and held for ransom each year in Phoenix alone.[31] And, as legal guidelines have tightened with the introduction of Operation Streamline in 2005, a rising number of migrants spend months in penal institutions in the United States before being deported back to their country of origin. Gatekeeper and the border policies that followed it have succeeded in creating a deterrent, but to achieve this deterrence migrants have suffered horribly through deportation, incarceration, kidnapping and death.

III. A HISTORY OF MEXICAN LABOR IN THE UNITED STATES

It is important to remember that over the last 100 years, the U.S. government and U.S. business interests have regularly invited Mexicans to work in the United States. At the beginning of the twentieth century, thousands of Mexicans moved north to the United States in one of the largest migrations in human history. The U.S. policy of restricting Asian immigration resulted in labor shortages in western states and created a strong demand for Mexican labor. From 1900 to 1930, in the southwestern states the Mexican population expanded from approximately 375,000 to 1,160,000.[32] Mexican workers were actively recruited by both the railroads and agricultural industry. By the 1920s, Mexicans comprised 75 percent of the 200,000 farmworkers in California and 85 percent of the 300,000 migrant farmworkers in Texas.[33]

From 1942 to 1964, in implementing the Bracero Program, the United States and Mexican governments collaborated to recruit workers from Mexico who were given temporary visas to work in agriculture and in building the railroads. Over the twenty-two years of this program, more than 4.5 million work contracts were granted and roughly 2 million Mexican workers were employed in the United States.[34] When the Bracero Program ended, the Immigration and Naturalization Act of 1965 introduced the H-2A and H-2B programs that gave temporary visas to foreign workers in those cases where the employer could show that no U.S. citizens were willing to accept employment at the prevailing wage.[35] However, rather than employing workers through the H-2A programs, most large agricultural employers opted for the less bureaucratic option of hiring undocumented workers.

The direct recruitment of undocumented Mexican workers by agricultural companies is a practice that continues into the twenty-first century. Bill Ong Hing, Professor of Law at the University of San Francisco, describes the practice of recruiting Mexican labor based on research from 2001:

> Organized groups of farm contractors travel to Mexican cities and towns, where they offer loans and work guarantees to convince potential farm workers to cross the border into the United States. The process involves well-organized networks of contractors and contractor's agents representing major U.S. agricultural companies. The headhunters are themselves often Mexicans who recruit in their own hometowns and farming communities, where earning the trust of eager farm hands is not difficult. One of the contractors' favorite tactics to attract workers is to offer them loans to help pay off debts, coupled with a pledge to find work for the person north of the border.[36]

In contemporary discussions of immigration, Mexican migrants are often depicted as lawbreakers that illegally enter the United States in search of employment. While it is true that undocumented workers enter the United States in violation of U.S. immigration law, the focus on their illegality conceals the important role that both the U.S. government and business interests have played in recruiting Mexican workers over the last 100 years. A careful examination of the history of Mexican labor in the United States, including the recruitment of Mexican labor in the 1920s, the twenty-two years of the Bracero Program, the H-2A visa program, and the ongoing recruitment of Mexican labor, demonstrates the central roles that U.S. government and industry have played in encouraging Mexican migration.

IV. COLLECTIVE RESPONSIBILITY

In this section, I use Larry May's account of collective responsibility to argue that the U.S. Congress is collectively responsible for failing to prevent the deaths of Mexican migrants. After a brief exegesis of May's account of collective responsibility, I use it to examine the issue of Mexican migration.

In *Sharing Responsibility*, May examines cases of collective inaction; that is, he looks at cases in which groups of people could have acted to prevent some harm but did not. As a central example, May imagines a case in which a child is screaming for help in the cold water a quarter of a mile from the shore. There is no life guard and none of the bystanders on the shore are capable of saving the child through individual action. Only collective action can save the child.[37] Cases like this involve what May refers to as a putative group; that is, although they are not initially a group, the people on the beach have the possibility of forming a group in order to mount an effective response to the crisis.

May argues that although individuals should not be held guilty for their collective inaction, they should feel shame.

> In the cases we have been considering, in which the failure of many people to act is involved, responsibility normally does not entail guilt. Shame, though, is directly related to a person's conception of herself or himself, rather than to explicit behavior (which is what guilt most commonly attaches to). Because shame concerns the self's identity, it is more appropriately felt than guilt when one's group fails to prevent a harm.[38]

Nevertheless, May argues that shame is a powerful moral feeling; and, although it does not place blame on particular individuals in the way that guilt does, May believes that shame can be a useful force in changing human behavior.

As a summary of his position, May offers the following:

> A putative group of people engages in the kind of inaction that warrants collective responsibility if (a) the members of the group fail to act to prevent a harm, the prevention of which would have required the coordinated actions of (some of) the members of the group; (b) it is plausible to think that the group could have developed a sufficient structure in time to allow the group to act collectively to prevent the harm; and (c) it is reasonable to think that the members of the group should have acted to prevent the harm rather than doing anything else, such as preventing other harms which they also could have prevented.[39]

Using May's theory as a guide, it seems appropriate to ask: Is there some putative group that should act to prevent the roughly 400 migrant deaths that take place every year as undocumented immigrants try to enter the United States? Is it reasonable to think that members of this group should have acted to prevent this harm as opposed to doing anything else?

Although humanitarian organizations (including No More Deaths and the Tucson Samaritans) have prevented a number of deaths each year by putting out water in the desert, they lack the ability to prevent the majority of the migrant deaths. In fact, there seems to be only one group that has the power to act in this case: the U.S. Congress. Members of Congress could form a coalition on this issue, pass a law and change the situation along the U.S. border to prevent a majority of these deaths. When a member of Congress is confronted with the stark reality of roughly 400 annual migrant deaths along the border, should he or she feel some degree of collective responsibility to prevent these harms? After all, a member of Congress is confronted by a number of different humanitarian crises each year: the civil wars in Syria and South Sudan, the thousands of people dying each year from malaria and AIDS, the suppression of human rights in various countries, et cetera. When compared to these other problems, in numerical terms, a mere 400 deaths annually might seem like an insignificant figure. Given that any member of Congress can only work on a limited number of projects, why should we think a member of Congress should feel obligated to build a coalition to address the problem of migrant deaths?

In thinking about which harms a putative group should feel collective responsibility to prevent, on May's account, putative groups are more responsible for preventing harms in cases where

1. the group has done something to bring about the harms;
2. the harms result from expectations raised by the group; and
3. the putative group is uniquely positioned to prevent the harms.

Let's consider each of these criteria in relation to the case at hand.

In the case of the 400 annual migrant deaths occurring along the U.S. border, did the U.S. Congress do something to bring about these harms? Yes. As we saw in the first section of this chapter, the passage of NAFTA resulted in a loss of more than 2 million agricultural jobs in Mexico in the first ten years of the agreement. Many of these displaced agricultural workers attempted to migrate to the United States and hundreds died trying to cross the border. In addition, as we saw in the second section, by implementing Operation Blockade and Operation Gatekeeper, the Border Patrol shifted the migration traffic from San Diego/Tijuana and El Paso/Juarez to the deserts of Arizona where migrants are much more likely to die. The combined effect of these two major policy changes increased the number of migrant deaths along the border by more than 500 percent.

Did the 400 annual migrant deaths result from expectations raised by the U.S. government? Yes. As we saw in the third section, during the Bracero Program (1942–1964), roughly 2 million Mexican workers were employed in the United States. And with the end of the Bracero Program, the government introduced the H-2A visa program to bring Mexican workers to the United States. Some regions of Mexico have been sending men to work in the United States for generations and the economies of these regions have been shaped by the continuing flow of remittances. Thus, U.S. policies have undoubtedly raised the expectations of Mexican workers that they could find work in the United States.

Is the U.S. Congress uniquely positioned to prevent the migrant deaths? Yes. The migrant deaths are occurring along the U.S. border in territory under the control of the United States. And, these deaths are occurring in large part due to the border policies and trade policies of the United States, policies that only the U.S. Congress has the power to change. Thus, it is clear that the U.S. Congress is uniquely positioned to prevent these deaths.

This situation fulfills all three of the criteria that May argues makes a putative group responsible for failing to prevent harms. Thus, on May's account, the members of the U.S. Congress are collectively responsible for their failure to do anything to prevent migrant deaths. Like the bystanders on the shore in May's central example, the members of Congress have the ability to form a group that would act to prevent the loss of innocent life; and, in their failure to act, they are collectively responsible for allowing these deaths to occur.

On May's account, feelings of guilt are appropriate in situations where an agent has acted individually or as a member of a group to cause some harm(s) for which he or she is morally responsible; but, feelings of shame are appropriate when an agent is a member of a putative group that has failed to act to prevent some harm(s). In what follows, I argue that in relation to Mexican migration the members of the U.S. Congress are best characterized as agents who failed to work together to prevent deaths of migrants, not as agents who caused the deaths of the migrants. Thus, in responding to the deaths of migrants, the appropriate response of members of Congress should not be feelings of guilt but feelings of shame.

There are three reasons to reject the strong claim that the members of the U.S. Congress are responsible for causing the deaths of the migrants. First, although NAFTA has resulted in significant job losses for Mexican agricultural workers, they still have a number of choices. Some decide to shift from corn production to crops that are more highly valued in the U.S. market, such as broccoli or cauliflower. Some move from rural areas to seek low-skill jobs in Mexico's cities. (In the ten years following NAFTA, total employment in Mexico's fast food industry increased by 527,285 jobs.)[40] Some decide to take a gamble with high returns and high risks by migrating to the United States. Anthropological research demonstrates that Mexican migrants are aware that they may die while attempting to cross the border.[41] Although NAFTA has limited their options, by choosing to engage in the risky behavior of border crossing, those who die bear some responsibility for their own deaths. Second, the United States has a legitimate interest in pursuing immigration control. In general, nation states have the right to enforce immigration restrictions. In this particular case, the United States has good reasons to restrict Mexican migration, both to protect the economic interests of low-skilled and marginalized workers (including inner-city African Americans, Native Americans living in poverty on reservations and Appalachian whites) and also to provide legal immigration opportunities to migrants in desperate circumstances (such as immigrants from South Sudan, Somali, or Syria). Thus, although Gatekeeper led to a rise in migrant deaths, by increasing deterrence, Gatekeeper has contributed to the U.S. government's legitimate interest in pursuing immigration restrictions. Third, for representatives who first joined the Congress in 1996 or later guilt seems like an inappropriate response, for these representatives played no role in passing the legislation that contributed to the death of migrants. Thus, for these three reasons, I reject the strong claim that the members of the U.S. Congress are responsible for causing the deaths of the Mexican migrants.

I argue for the weaker claim that the members of Congress should feel shame for their failure to prevent the migrant deaths. Although the U.S. Congress didn't cause the migrant deaths, the actions of the U.S. Congress created a social context in which migrant deaths are more likely to occur. NAFTA contributed to increases in Mexican migration; Gatekeeper shifted migration routes to more dangerous areas. As hundreds of migrants died each summer, Congress has not acted to prevent this humanitarian crisis in the border region. When faced with the gruesome reality of death by dehydration that regularly occurs in the border region, the members of Congress should feel shame for belonging to a group that passed legislation that contributed to this crisis and has done nothing to prevent the deaths that regularly occur. Such feelings of shame should motivate the members of Congress to form a coalition to draft and pass legislation to address the humanitarian crisis of migrant deaths.

V. POLICY RECOMMENDATIONS

Having shown that the U.S. Congress is collectively responsible for failing to prevent the migrant deaths that have resulted from border and trade policies, in this section I present two policy recommendations that would reduce Mexican migration and the number of migrant deaths along the border. The two policy proposals I advocate are (1) the construction of lifesaving towers along the U.S.–Mexico border and (2) the elimination of agricultural subsidies in the United States. These proposals could be passed individually or included as part of a broader program of comprehensive immigration reform.

Proposal 1: Lifesaving Towers

Like front-line troops in a war zone, Border Patrol agents are directly exposed to the horrific consequences of border and trade policies. With hundreds of migrants dying of dehydration each year, the deserts of the border region are increasingly littered with corpses. When bodies are recovered, it is usually Border Patrol agents who discover the bodies. In his detailed reporting on the Yuma 14, a group of fourteen migrants who died in the Yuma sector while trying to cross the desert, Luis Alberto Urrea notes that Border Patrol agents spoke of the tragedy indirectly: "They (Border Patrol agents) refer to the catastrophe in obtuse terms—not 'Yuma 14' for them. They still say 'the thing that happened' when they speak of it. As if it were too sad to name."[42] In response to the tragedy of the Yuma 14, Border Patrol agents in the Yuma sector developed a technical solution to minimize the number of migrant deaths. In some of the remote regions of the desert where migrants were most likely to die, Border Patrol agents have erected lifesaving towers. The towers are 30 feet tall, with reflectors and beacons. The beacons flash every 10 seconds. The towers are thus visible at any time of the day. Signs (in both English and Spanish) warn: "Attention! You cannot walk to safety from this point! You are in danger of Dying if you do not summon help! If you need help, push red button. US Border Patrol will arrive in 1 Hour. Do not leave this location."[43] A few months after they were first deployed, the towers in Yuma sector proved their utility. When three migrants pushed the red button, the Border Patrol agents from Welton station rescued them and seventeen others they had left behind in the desert. Initial observations indicate that the death rates for migrants have dropped significantly in Border Patrol sectors where lifesaving towers have been deployed. This effort to save lives has been a grassroots effort implemented through the collective efforts of Border Patrol agents. "The towers are built, raised, maintained, and paid for out of pocket by those bleeding heart liberals, the Border Patrol agents themselves."[44] Although proimmigration activists have often depicted Border Patrol agents as cruel enforcers of inhumane policies, as Urrea's research demonstrates, the great majority of Border Patrol agents express sympathy for the plight of economic migrants.

One obvious solution to the ongoing tragedy of migrant deaths is that the Border Patrol should follow the example of agents in the Yuma sector and construct lifesaving towers throughout the most dangerous regions. Each tower costs only $6,000, so the deployment of 10,000 towers throughout the border region would cost $60 million. By comparison, the border wall being constructed along the U.S.–Mexico border costs an average of $1.3 million per mile.[45] So the cost of building 10,000 lifesaving towers is equal to the price of building roughly 46 miles of border wall. Given its potential to save hundreds of migrants from death each year, this solution should be implemented immediately.

Proposal 2: The Elimination of U.S. Agricultural Subsidies

Free trade agreements can lead to significant economic growth and prosperity for the countries involved, but for all participants to benefit, free trade must be implemented in a manner that is fair and just. Agricultural subsidies in first world nations privilege special interests groups in the first world at the expense of large populations of agricultural workers in developing nations. As a result, agricultural subsidies can significantly undermine the benefits that developing nations can gain from free trade agreements. The issue of first world agricultural subsidies has become a sticking point in free trade agreements. In the July 2008 meeting of the World Trade Organization negotiations broke down when developing nations objected to agricultural subsidies in the United States and European Union.[46] In a case that is similar to the dumping of subsidized corn in Mexico, European agricultural subsidies have done significant harm to farmers in Africa.

> It does not take a professional economist to realize that African economies are predominately agricultural and that European (and American) policies are increasingly suffocating and asphyxiating the continent. These subsidies leave African markets flooded with cheap European and American goods. Chicken, tomato paste, onions, fruit and vegetables and other agricultural products from the EU now flood the African market. African farmers . . . have little option but to "employ their feet" in this desperate attempt at survival.[47]

In light of NAFTA's devastating effects on agricultural employment in Mexico, the agricultural components of NAFTA need to be reconsidered. In the first ten years of NAFTA alone, more than 2 million jobs were lost in Mexico's agricultural sector. A number of factors make it difficult for small Mexican farmers to compete with agricultural imports from the United States: they have limited access to the latest technology for seeds and fertilizer, warehousing and transportation costs are high due to poor infrastructure, they have difficulty attaining financial support from banking institutions, and most small Mexican farmers do not have access to tractors. Due to sizable agricultural subsidies provided by the U.S. government, agricultural products are sold in Mexico at prices well below their cost of production. To take the example of corn, the primary agricultural product for a majority of farmers in Mexico, due to government subsidies, U.S. corn is sold at a price 30 percent below what it actually cost the U.S. farmers to produce the corn.[48] In the years following NAFTA, Mexican agricultural subsidies and technical assistance were reduced 75 percent (subsidies and assistance totaled $2 billion in 1994 and only $500 million in 2000).[49] As a result of the continuation of sizable agricultural subsidies in the United States and the reduction of agricultural subsidies in Mexico, uncompetitive agricultural jobs are created in the United States at significant cost to the U.S. taxpayer and agricultural jobs are destroyed in Mexico as farmers fail to compete with subsidized imports. These policies are clearly unjust.

In order to rectify these inequities, I advocate a renegotiation of NAFTA's agricultural provisions that would require the gradual elimination of U.S. agricultural subsidies over a five-year period. Agricultural subsidies create more agricultural jobs in the United States that are often filled by undocumented Mexican workers. Thus, by artificially inflating the size of U.S. agricultural operations, U.S. agricultural subsidies create "pull" factors that encourage Mexican workers to migrate to the United States.

To take one example, the production of sugar in the Unites States has been uncompetitive with global prices for the last fifty years. Yet, government price supports have guaranteed U.S. farmers prices for sugar that have at times been three times higher than the world market price.[50] As a result of these subsidies, a totally uncompetitive U.S. sugar industry continues to employ roughly 17,000 workers.[51] The elimination of subsidies for sugar would eliminate the U.S. sugar industry and reduce the "pull" factor for undocumented workers by eliminating 17,000 low-paying jobs in the sugar industry. And, the elimination of sugar subsidies would save American consumers roughly $1.4 billion annually.[52] While government subsidies have artificially preserved the uncompetitive U.S. sugar industry, in Mexico, in the post-NAFTA period, free market reforms led to massive jobs losses in the sugar industry. In 1998, 420,085 workers were employed in Mexico's sugar industry, but by 2006, only 129,476 workers were so employed.[53]

The elimination of U.S. agricultural subsidies would produce four significant results. First, it would rectify current inequities in NAFTA to offer Mexico a trade deal that is not merely free trade but also fair trade. It would put an end to a biased system that has created unfair advantages for U.S. farmers at the expense of desperately poor Mexican farmers. Second, this policy change would benefit the American taxpayer, who would no longer be forced to waste money on agricultural subsidies, and the American consumer, who would no longer pay higher prices for agricultural products supported by price controls. Third, by creating a level playing field, this policy change might lead to an expansion of the agricultural sector in Mexico and the creation of more agricultural jobs in Mexico. At the very least, this policy change might prevent future job losses in Mexico's agricultural sector. Fourth, the elimination of agricultural subsidies would cause a reduction in the size of the agricultural workforce in the United States as uncompetitive production (e.g., the U.S. sugar industry) was reduced. The third and fourth results listed above address the underlying "push" and "pull" forces that produce economic migration. By producing more jobs in Mexico's agricultural sector and reducing the number of low-paying agricultural jobs generally filled by Mexican laborers in the United States, the elimination of agricultural subsidies in the United States would reduce Mexican migration.

VI. CONCLUSION

In this chapter, I began by examining some basic facts about Mexican migration. First, I showed that NAFTA had devastating consequences for Mexican farmers that produced an increase in Mexican migration. Second, I examined the long history of U.S. government and business efforts to bring Mexican labor to the United States. Third, I demonstrated that Operation Gatekeeper and subsequent changes to border policy have quintupled the number of migrant deaths along the border, resulting in more than 5,000 migrant deaths in the last fifteen years. In light of these facts, I used the ideas of philosopher Larry May to argue that members of the U.S. Congress are collectively responsible for failing to act to prevent the migrant deaths that have resulted from the combined effects of our trade and border policies. And, I presented two policy proposals that would reduce Mexican migration and migrant deaths along the border: the construction of lifesaving towers in the border region and the elimination of U.S. agricultural subsidies. The crisis along the border is an annual event. With the arrival of each summer in the southwest, daytime temperatures hit triple digits, and scores of men, women and children, pursuing dreams for a better life, find death instead. How many more summers will pass before we can bring this tragedy to an end?

NOTES

1. Charles Wheelan, *Naked Economics: Undressing the Dismal Science* (New York: W. W. Norton & Company, 2010), 274–75.
2. Ernesto Zedillo Ponce de Leon, "NAFTA at 20," *Americas Quarterly* (Winter 2014): 28.
3. Juan Rivera and Scott Whiteford, "Mexican Agriculture and NAFTA—Prospects for Change," in *NAFTA and the Campesinos: The Impact of NAFTA on Small-Scale Agricultural Producers in Mexico and the Prospects for Change*, ed. Juan Rivera, Scott Whiteford, and Manuel Chavez (Scranton: University of Scranton Press, 2008), xv.
4. Rivera and Whiteford, "Mexican Agriculture," xv–xvi.
5. Judith Adler-Hellman, *The World of Mexican Migrants: The Rock and the Hard Place* (New York: New Press, 2008), 17–19.
6. Juan Rivera, "Multinational Agribusiness and Small Corn Producers in Rural Mexico: New Alternatives for Agricultural Development," in *NAFTA and the Campesinos: The Impact of NAFTA on Small-Scale Agricultural Producers in Mexico and the Prospects for Change*, ed. Juan Rivera, Scott Whiteford, and Manuel Chavez (Scranton: University of Scranton Press, 2008), 89.
7. Rivera, "Multinational Agribusiness," 99.
8. Manuel Angel Gomez Cruz and Rita Schwentesius Rindermann, "NAFTA's Impact on Mexican Agriculture: An Overview," in *NAFTA and the Campesinos: The Impact of NAFTA on Small-Scale Agricultural Producers in Mexico and the Prospects for Change*, ed. Juan Rivera, Scott Whiteford, and Manuel Chavez (Scranton: University of Scranton Press, 2008), 4–5.
9. Rivera, "Multinational Agribusiness," 90.
10. Cruz and Rindermann, "NAFTA's Impact," 4–5. See also, Rivera, "Multinational Agribusiness," 91.
11. Cruz and Rindermann, "NAFTA's Impact," 6–7.
12. Ibid.
13. Ana Carrigan, "Chiapas, The First Postmodern Revolution," in *Our Word is Our Weapon: Selected Writings, Subcomandante Insurgente Marcos*, ed. Juana Ponce de Leon (New York: Seven Stories Press, 2001), 417–43.
14. Rivera, "Multinational Agribusiness," 90.
15 Ibid., 101.
16. Cruz and Rindermann, "NAFTA's Impact," 11.
17. Ronald Takaki, *A Different Mirror: A History of Multicultural America* (New York: Back Bay Books, 2008), 427.

18. Joseph Nevins, *Operation Gatekeeper and Beyond: The War on "Illegals" and the Remaking of the U.S.–Mexico Boundary* (New York: Routledge, 2010), 140.
19. Ibid., 227.
20. Ibid., 105.
21. Ibid.
22 Ibid., 145.
23. Ken Ellingwood, *Hard Line: Life and Death on the U.S.–Mexico Border* (New York: Random House, 2005), 32.
24. Ibid., 33.
25. Ibid., 68.
26. Ruth Gomberg-Munoz, *Labor and Legality: An Ethnography of a Mexican Immigrant Network* (New York: Oxford University Press, 2011), 35.
27. Nevins, *Operation Gatekeeper*, 174.
28. No author listed, "The US–Mexico Border: Secure Enough," *The Economist*, June 22, 2013, 31.
29. *The Economist*, June 22, 2013, 32.
30. Ibid.
31. Terry Greene Sterline, *Illegal: Life and Death in Arizona's Immigration War Zone* (Guilford: Lyons Press, 2010), 43.
32. Takaki, *A Different Mirror*, 295.
33. Ibid., 297.
34. Ronald L. Mize and Alicia Swords. *Consuming Mexican Labor: From the Bracero Program to NAFTA* (Toronto: University of Toronto Press, 2011), 3.
35. Ibid., 95.
36. Bill Ong Hing, *Ethical Borders: NAFTA, Globalization, and Mexican Migration* (Philadelphia: Temple University Press, 2010), 121.
37. Larry May, *Sharing Responsibility* (Chicago: University of Chicago Press, 1992), 110.
38. Ibid., 120.
39. Ibid., 117.
40. Cruz and Rindermann, "NAFTA's Impact," 11.
41. Seth M. Holmes, *Fresh Fruit, Broken Bodies: Migrant Farmworkers in the United States* (Los Angeles: University of California Press, 2013), 8.
42. Luis Alberto Urrea, *The Devil's Highway* (New York: Back Bay Books, 2005), 212.
43. Urrea, *The Devil's Highway*, 213.
44. Ibid.
45. Ong Hing, *Ethical Borders,* 46.
46. Wheelan, *Naked Economics*, 281.
47. Tongkeh Fowale, quoted in Ong Hing, *Ethical Borders*, 94.
48. Ong Hing, *Ethical Borders,* 14.
49. Ibid., 46.
50. Gerardo Otero and Cornelia Butler-Flora, "Sweet Protectionism: State Policy and Employment in the Sugar Industries of the NAFTA Countries," in *NAFTA and the Campesinos: The Impact of NAFTA on Small-Scale Agricultural Producers in Mexico*

and the Prospects for Change, ed. Juan Rivera, Scott Whiteford, and Manuel Chavez (Scranton: University of Scranton Press, 2008), 70.

51 Ibid., 85.

52. Ibid., 70.

53. Ibid., 77.

Part III

FOOD AND GENDER

Chapter 8

Food is a Feminist Issue

Lori Watson

If asked to list the primary sources and effects of gender injustice, few people (at least in the Western world) are likely to include the ethical issues surrounding food near the top of the list.[1] Indeed, if one is asked to name the priorities of feminism—the primary lens through which gender justice is theorized—sexual violence, sex discrimination and harassment, the underrepresentation of women in positions of power, both in the private and public sectors, as well as the unequal burdens of childcare and domestic labor are likely to be considered front and center. As evidence, consider that many of the primary texts of feminist scholarship (in the West) make no mention of food.[2] Martha Nussbaum's work on women and human development is an important exception; however, her focus is not on food justice per se.[3]

None of this is to say that considerations of food justice are not seen as critical to human rights; indeed, they are. Poverty, hunger, lack of nutrition and basic resources for survival are primary forms of inequality and injustice that scholars, citizens, governments, international political institutions and NGOs emphasize require urgent action. And while the United Nations and various NGOs[4] carefully attend to the connection between food justice and gender justice, it is commonplace in public discourse to see them as separate issues. In this chapter, I want to explore food justice through the lens of gender. In particular, I argue that food justice is a central issue of gender justice and that by creating a more gender just world, both globally and locally, we will make great strides in securing food justice.

I. JUSTICE

For many, concerns about justice are primarily concerns about fairness. More specifically, many political theorists think one primary element of justice has to do with distributive fairness. In other words, if we want to evaluate the justness of a given society, or develop an ideal of justice, many think we should focus our attention on how resources are distributed among the members of a given society, or even globally among all persons. As such, theories of distributive justice focus on the ways in which income, wealth, property, goods, resources, etc. are allocated within their scope of concern (a nation-state or globally).[5] Theories of distributive justice, then, offer an account of how to morally evaluate the economic framework of a given society or the global society, comprised of all nation-states, and the way resources are distributed within or among them. Not all theories of distributive justice focus exclusively on material resources and their distribution. For example, John Rawls includes the fair value of political liberties, and fair equality of opportunity, the social basis of self-respect, as well as the benefits and burdens of social cooperation among other goods as the primary goods with which distributive justice is concerned.[6]

Insofar as many theorists of distributive justice are concerned with resources, in particular, they disagree over what resources are significant for assessing the fairness of a distributive scheme: some favor income, others wealth, others utility, others welfare (understood as well-being), to name a few. They also disagree over what set of persons is relevant to assessing a just distribution: individual persons or groups. And significantly, they disagree over what principles of distribution constitute a fair distribution: principles that entail distribution based on desert, merit, or strict egalitarianism, or some other principle, for example.

Feminist critics have pointed out that much of contemporary writing on distributive justice has tended to focus too narrowly on the distribution of material resources (however defined) and offering suggestions for rectifying unequal distributions without sufficiently attending to the social patterns of exclusion and oppression, especially gender oppression, that underwrite patterns of unfair distribution and other forms of injustice.[7] For example, feminist authors such as Iris Marion Young, Nancy Fraser, Elizabeth Anderson and Martha Nussbaum argue that we ought to broaden our conception of justice beyond a narrow focus on distribution of resources per se and look to the social institutions and patterns, which shape our lives more broadly, as matters of fundamental justice.[8] Oppression and domination are two primary sources of injustice, where oppression is understood as "the institutional constraint on self-development, and domination, the institutional constraint on self-determination."[9] And while oppression may involve "material deprivation or maldistribution" it crucially involves much more, as does domination.[10] Nancy Fraser urges us to see the centrality of recognition, especially equal recognition for subordinated social groups, as a central element of justice alongside redistribution (of, say, wealth and material resources).[11]

Whereas Martha Nussbaum has developed a "capabilities account" of justice, expanding and going beyond the work of Amartya Sen, that urges us to evaluate the quality of life of all persons as defined by their capabilities. As such, a capabilities account goes far beyond traditional quality of life metrics that focused on Gross Domestic Product (GDP) as a measure of quality of life.[12]

Elizabeth Anderson, in critiquing much of the mainstream literature on distributive justice, and egalitarian principles of distributive justice in particular, reminds us that "the point of equality" is to oppose oppression and hierarchy that aims to "justify" inequalities in the first place. Egalitarians are committed to the equal moral worth of all persons, from which two claims follow, according to Anderson. "Negatively, the claim repudiates distinctions of moral worth based on birth or social identity—on family membership, inherited social status, race, ethnicity, gender or genes. . . . Positively, the claim asserts that all competent adults are equally moral agents; everyone equally has the power to develop and exercise moral responsibility, to cooperate with others according to principles of justice, to shape and fulfill a conception of the good."[13] As such, egalitarians should seek to "abolish oppression" and "create a social order in which persons stand in a relation of equality."[14]

Anderson calls her view a relational theory of equality and contrasts it with a purely distributive theory of equality. A purely distributive theory of equality understands equality as a function of the patterns of the distribution of resources (or welfare, or utility, or whatever metric is favored). Under such a theory, persons will be said to be equal just in case the distribution of the relevant good they receive is equal according to the theory's preferred metric. In contrast, democratic egalitarians (those who endorse a relational theory of equality) are primarily concerned with the "relationships within which goods are distributed, not only with the distribution of goods themselves."[15] Hierarchical social identities such as those based on racial categorization, sex and gender distinctions, sexuality, ethnicity and religion, among many others, constitute the social fabric within which patterns of distribution occur. Aiming to eliminate such hierarchical social distinctions is central to equality as a moral value, as such distinctions are antithetical to the equal moral worth of all persons. However, eliminating such hierarchies is also instrumentally valuable for creating sustainable and just patterns of economic distribution.

This relational theory of equality is superior to a purely distributive theory of equality in several respects: first, it makes it clear why some goods are much more significant than others when assessing a society's economic framework and the patterns of distribution. Inequality with respect to basic necessities of life, opportunities for political participation, education and meaningful employment serve to reinforce the subordinate status of some relative to others, and often on the basis of group membership (e.g., race and gender). As such, these kinds of inequalities are a matter of fundamental justice, and require urgent action. Second, it makes clear that certain kinds of interventions (whether by particular nation-states or the United Nations) are superior to others.

So, for example, some have argued that we presently have enough food to solve world hunger, and the main problem is inefficiency in distribution of food.[16] One might read Peter Singer's very influential article "Famine, Affluence, and Morality," and his more recent book *The Life You Can Save*, as emphasizing the importance of redistribution of wealth as central to solving the problem of world hunger.[17] As such, one may legitimately complain that Singer's proposal suffers from some serious flaws: first, it frames the issue in terms of transferring goods and resources from some people to others. And, while no doubt those who are flush with resources have some duty to aid those with scarce resources, this way of addressing the issue omits or elides some very salient moral features of the situation in which many people in the world are lacking basic resources for survival. Focusing on the problem of world hunger as merely a distribution problem (not getting the food to the people who need it) misses the underlying patterns of inequality that structure and sustain unequal distributions. This is not to say that better patterns of distribution, and getting more food to needy people, isn't a worthy goal. However, addressing the structural and systemic obstacles to equal standing that prevent the emergence and sustainability of patterns of equal distribution is a central goal, from the point of view of relational egalitarians.

A concrete example of this point: the United Nations has recognized that gender equality is a necessary means of eliminating distributional inequalities, rather than simply aiming for better distribution as a means to more gender equality, in its Millennium Development Goals, and a variety of other contexts.[18] In describing the shift from a "woman in development" (WID) policy to a "gender and development" (GAD) policy, a UN report notes that although WID had short-term benefits for things such as health, income and resources, it did little to address inequalities in relationships or the structural components that enforce such inequalities.[19] The shift to a GAD policy had as its goal "people centered development" and thus the removal of social, political and economic inequalities between men and women.[20] This policy, along with a further policy of focusing on women's empowerment, highlights the importance of just the kind of relational equality we have been concerned with as central to creating sustainable and ongoing patterns of just distribution.

Eradicating extreme poverty and hunger is the first of the eight Millennium Development Goals the United Nations articulated. Gender equality is a necessary condition for achieving that goal. Moreover, promoting gender equality in its own right and empowering women is the third of the Millennium Development Goals; this goal recognizes that only through gender equality will our most important developmental goals be achieved and our aspirations of living in a more just world become closer to reality. Enhancing women's and girls' food security, empowering them to own and cultivate land on equitable terms, and ensuring their access to clean and safe water are central issues of gender justice, globally. I now turn to these topics.

II. FOOD SECURITY

Food security is a pressing issue of gender justice. At the 1996 World Food Summit, food security was characterized as existing "when all people at all times have access to sufficient, safe, nutritious food to maintain a healthy and active life."[21] Food security is defined in terms of three variables:

- food availability: that there is an adequate and consistent supply of nutritious food;
- food access: that individuals have adequate resources to obtain nutritious food; and
- food use: that individuals have knowledge regarding the proper use of food (nutrition and care), and that there is adequate water and sanitation.[22]

On each of the three variables, women and girls are disproportionality food insecure. Globally, 60 percent of undernourished persons are women and girls.[23] Women and children constitute a majority of those living in extreme poverty. According to many reports, women and children make up two-thirds of the world's persons living in extreme poverty.[24] Though some have questioned the quality of the data from which this conclusion is drawn, it is clear that women and girls are more vulnerable to poverty and more vulnerable due to poverty.[25] Such poverty undermines food security for men and women, boys and girls. However, women and girls suffer food insecurity disproportionately within their communities and families. For example, within households women and girls often receive less than their equal share of resources, including food, as a result of custom and the tradition of valuing boys over girls and thus seeing boys' nutritional needs as primary. Further supporting this pattern of unequal access and sharing of food are social expectations of feminine self-sacrifice and caring for others, such that women and girls are often taught to attend to others' needs before they attend to their own.

Gender inequality is both a cause and consequence of such food insecurity. The undervaluing of women and girls, relative to men and boys, in many societies creates and sustains social patterns in which women (and girls) have less access to resources, remunerated labor or comparable remuneration, education and educational opportunities. These patterns of gender-based inequality contribute to a weakened bargaining position and value within families.

As a result, women and girls have little to no say in household matters. Further, feeding and caregiving practices tend to favor boys and men, which leads to food and nutrition insecurity for women and girls.[26] In many parts of South Asia, for example, women eat only after the men and boy children have eaten.[27] Part of the explanation for such unequal practices is that women and girls adapt their preferences to their circumstances, and so do not claim or seem to even want for themselves equity. The lesser status of women and girls within families and communities means that they fail to have access to available resources on an equitable basis, including, importantly, food.

Due to the subordinate status of women and girls, they are not only excluded from equal access to education and employment opportunities, but also the gender-based work they contribute to the family's maintenance is often not recognized as work and is comparably undervalued. Women and girls do an overwhelmingly disproportionate amount of care work, much of which is unpaid. In fact, roughly two-thirds of women in developing countries do unpaid work in the home or work in the informal sector.[28] Women are constrained by laws and customs in their participation in the labor force, often failing to earn adequate income that would enable them to purchase basic goods, including adequate amounts of food.

But, also, the consequences of food insecurity for women and girls deepen gender inequality along various other social dimensions: health and nutrition of women and girls are threatened and maternal health, as well as the health of children, is severely impacted as a result of gender inequality. Women and girls are more likely to be malnourished than men; they suffer greater rates of anemia, iodine deficiency, vitamin A deficiency, and are more likely to be underweight.[29] The malnutrition of women has a cascade of other consequences: it leaves women susceptible to infections, leaves them vulnerable in childbirth, undermines their productivity, undermines their ability to earn income and undermines their ability to provide care for their families.[30] When women are malnourished, their children are more likely to be malnourished.[31] Recent research has demonstrated that malnutrition is associated with 60 percent of the deaths of children five and younger, and there is a strong correlation between a mother's poor nutrition and her child's malnutrition.[32] Food insecurity not only impacts the health and safety of women, but it also hampers their ability to produce, prepare, process, distribute and market food, all of which are central to the health and vitality of communities.[33]

Rising food prices, changing patterns of food production, as well as droughts, climate change, natural disasters and armed conflict are significant threats to food security for vulnerable populations, especially in rural areas. Women are disproportionately impacted when humanitarian crises evolve due to the aforementioned causes. Women often have less access to humanitarian aid during such crises, and especially with ongoing armed conflict.[34] For example, according to the Food and Agriculture Organization (FAO) of the United Nations, in many places in the world, women's access to information about relief, relief networks and transportation is limited. Further, since women are less likely to have control over, or even access to, adequate food or food production necessities such as land, animals, or tools, their ability to recover from disasters is substantially hampered.[35] The reality of climate change and the increased patterns of extreme weather mean that women and children are likely to suffer disproportionately due to the effects of these crises, including increased food insecurity. In the short term, when specific disasters strike, the FAO recommends that gender-sensitive relief programs can help to forestall widespread malnutrition. Further, the recovery time is more likely to be faster and more widespread if the relief program is gender sensitive.[36] Such gender-sensitive interventions include evaluating the gender-differential impacts of disasters, collecting sex-disaggregated data on the basic elements of food security (access and availability, especially) so as to create empirically informed programs to address food insecurity for women and children in crises contexts, including women in key decision-making roles in planning relief efforts, and using women's specific knowledge as a resource in such planning.[37] Longer-term interventions aimed at reducing the gender-based impacts of climate change and weather-related disasters recommended by the FAO include creating safety nets that specifically target the food security of women and children in the context of a crisis, whether weather related or due to armed conflict, or some other humanitarian crisis; developing labor-saving strategies for reducing women and children's burdens in food production and water gathering to create more time for activities that would generate income; and making sure that the technologies used in response to crisis are sensitive to women's needs and capabilities.[38]

What is clear is that achieving food security for all people requires attending to the gendered patterns of inequality that create and sustain women's vulnerability and developing concrete policy strategies aimed at undoing patterns of gender hierarchy that exclude women from full and equal participation in all aspects of civil society. Aiming to eradicate patterns of gender inequality that sustain and reinforce women's subordinate status relative to men entails at least the following:

- Elimination of laws that prohibit or restrict women's ability to own land (see section III for further discussion of this issue).
- Elimination, or at minimum shifting, of norms that exclude women's access to land through inheritance.
- Increasing women's access to, and opportunities for, education, which will include among other things making access to clean safe water and toilets (see section IV for further discussion of this issue).

- Addressing the social patterns of behavior that reinforce gendered patterns of caregiving work. This may include both making men responsible for care giving (in all its forms, from children to the elderly) and efforts to enhance the status of caregiving work as work.[39]
- Active inclusion of women in local government, and inclusion of women in decision-making processes within their communities.
- Increasing women's participation in the labor force, opening up employment opportunities for women, and securing equal remuneration for paid work.
- Legal and social interventions aimed at eliminating gender-based discrimination.
- Increasing women and girl-children's status within families, so as to eliminate unequal food distribution along gender lines.

Addressing such gender inequality is necessary for achieving food security for everyone: "Gender equality can make a substantial contribution to a country's economic growth, and it is the single most important determinant of food security."[40]

III. AGRICULTURAL LABOR

Women play a significant role in food production, globally. For example, in Africa women produce more than 70 percent of the food.[41] In Asia, women produce 60 percent of the food, and likely more than that but exact numbers are elusive due to poor data specifically on women's agriculture labor.[42] One consequence of an increasingly global economy and economic development in various nation-states is that men are exiting agricultural forms of labor at increased rates. Men are more likely to move to urban centers for nonagrarian work, and women are excluded from many forms of labor, leaving them relegated to domestic and agricultural labor. Thus, women largely remain in the agrarian sector of the economy (both formal and informal) as their primary form of employment. Moreover, the global ratio of women to men as food producers is steadily increasing. For example, in 2008, 63 percent of female workers in Africa were dependent on agriculture for their livelihood, compared to 48 percent of male workers.[43] And in India, 65 percent of women workers are employed in agriculture, compared to 49 percent of men workers.[44] Eighty-three percent of Indian rural female workers are in the agriculture sector.[45] These facts have led some to claim there is a "feminization of agriculture."[46] Women engage in subsistence agriculture at higher rate than do men, and their contributions to food production are often not counted in official recordings of agricultural labor, thus their actual participation and contribution are likely undercounted.[47] Thus, it is clear that women's role in agricultural labor is a key factor in food security for everyone, as they are increasingly the primary food producers globally. Moreover, as they bear primary responsibility for food preparation and meeting the nutritional needs of the family, including importantly children, their role in ensuring food security for the family or community is of vital importance to all.

Yet, even where women are engaged in agricultural labor in greater numbers, they continue to be relegated to lower-status forms of labor and their contributions are undervalued relative to that of men. For example, in China, men take ruminants to the mountains to graze while women raise pigs and chickens.[48] Women tend to be planters of seed, while men plow and engage with the market.[49] The majority of farmers in the developing world work on small farms, including the majority of women engaged in agricultural labor.[50] Moreover, women are often excluded, formally through law or informally through custom and discrimination, from access to land, financial services (credit), support services (farming support) and markets. In the Congo, Tanzania, Nepal, Indonesia and Vietnam, the percentage of women owning land as a percentage of all landowners is less than 10 percent, in India and Malaysia it is less than 15 percent.[51]

The lack of land ownership for women is a significant threat to food security. Women depend on land both for sustenance and for accessing other important goods and services. Credit, for example, is difficult to obtain without land.[52] The primary obstacle for land ownership for women includes laws of inheritance, which often exclude women from inheritance rights to property. And even if formal law does not explicitly exclude women from inheritance rights, custom often does. For example, in the People's Republic of China legal reforms have granted women equal formal rights to land ownership, however custom dictates that sons and not daughters or widows are the rightful inheritors of land. Women hold only 7 percent of land certificates, and husbands and wives together hold another 5 percent, the rest are in men's names only.[53] Similarly, in India, formal laws have given women equal rights to their "natal family assets," yet women rarely inherit property due to customary norms.[54]

Land access and ownership are central to women's equality and to food security. While such access and ownership can and does improve women's economic lives, it also is a significant source for women's empowerment and social position within the family. Women's ability to hold assets has proven to improve their power to make household decisions, to increase their capacity to make autonomous choices about their reproductive lives and control the number of children they bear, and to lead to a substantial reduction in their experience of domestic violence.[55] Women's ability to own land also leads to an improvement in the welfare of their children and a reduction in child malnourishment. Further, land ownership has increased women's decision-making ability in the community, leading to increased status as citizens and greater political participation.[56] Equal access to land is essential to increasing women's standing within their communities and households, but is also significant for increasing food production, a key element of food security. Even more, where women are not food producers themselves, landownership can be used as collateral to secure loans to start small businesses and earn income, which in turn increases their ability to purchase the food they and their families need.

Not only do women farmers face inequality in land ownership, they lack access to many of the crucial supports to effective farming such as credit, fertilizers, water, marketing and technical information.[57] Further, women have limited access to tools necessary to cultivate and maintain their crops, due to limited resources and gender biases.[58] Women's role in food production is severely undermined by the lack of access to the necessary recourses for food production, and the result is greater food insecurity not only for women, but also for us all because, as noted earlier, women are increasingly the primary food producers in the global food market.

Gender equality is central to food productivity. According to the FAO, if women's access to productive resources were equal to that of men's, their farm's yield would increase by 20 to 30 percent. This would increase the agricultural output of developing nations by 2.5 to 4 percent and could go a long way toward feeding the hungry. The FAO estimates that gender equity in access to productive resources could reduce world hunger by 12 to 17 percent.[59] In order to overcome the obstacles for food security created and maintained by gender inequality, Bina Agarwal (Director and Professor of Economics at the Institute of Economic Growth, in Delhi, and member of the UN Committee for Development Policy) offers the following recommendations:

- "Recognizing women as farmers and not simply as farm helpers.
- Improving women's direct access to land and providing them with security of tenure.
- Increasing women's direct access to production credit, agricultural inputs, technology, technical information on improved agricultural practices, and marketing outlets.
- Directing more agricultural research and development to crops women cultivate and based on a better understanding of women's farming systems.
- Institutional innovations, in particular promoting a group approach to farm investment and investing."[60]

Each of these recommendations requires improving the status of women in their societies, specifically by undermining gendered hierarchies reflected in law and custom that serve to maintain the subordinate status of women and girls. The laws of inheritance need to be effectively changed, as well as the customs supporting male preference.

Microfinance programs are one key developmental tool for empowering those living in poverty. Broadly, microfinance programs are financial programs for the poor, which are outside of traditional banking services.[61] Microloans are loan programs available through various NGOs and microfinance institutions; often these programs target women specifically and loan only to women. Microfinance programs are potentially a useful tool in combating the limitations women face as food producers, due to the kinds of limitations regarding access to land, credit, tools and other resources mentioned earlier. These gender-centered programs emphasize the importance of including and empowering women economically as a key constituent of economic development for a given community. And while such programs are a central tool in combating poverty, there are still important gender barriers that need to be addressed. For example, while women may have access to loans to start businesses or develop small businesses, including farming, through microfinance programs, this can lead to empowerment of women within their household, often girl-children in the family are expected or required to undertake extra work in order that the adult women can engage in their business endeavors.[62] Moreover, men may still control the loans. And men may borrow through women for their own businesses. The evidence on whether microfinance programs have empowered women is thus mixed.[63] There is some evidence that men benefit more than women through microfinance, even when women are the primary recipients of the loans. For example, a qualitative study of microfinance loans was done in the early 1990s, roughly ten years after such programs had begun targeting women. The study looked at 275 loans, financed by four different organizations. It revealed that in 43 percent of the loans, women had either limited or partial control of the use of the fund, or no control at all. Another study in Andhra Pradesh (in India) revealed that the funds from 67 percent of women's loans were not invested in assets or businesses controlled by the women themselves, but, instead, by their husbands.[64] Moreover, women often end up with a heavier burden in repaying those loans. The same study in Andhra Pradesh found that in roughly 85 percent of the aforementioned cases, women had to engage in extra wage labor to make the repayments.[65] Increased support for women, through better training programs and education, is minimally necessary to increase the effectiveness of microfinance programs for alleviating gender-based poverty and empowering women.

The mixed evidence of the effectiveness of microfinance programs further highlights that efforts at better distribution of resources (money) or redistribution of such resources alone is insufficient for either evaluating women's status or for solving the food security issue. While it is true that better access to resources such as land, money, tools and other food production necessities is central to women's equality and improving food security, the underlying cultural and social norms that situate women as men's subordinates are foundational for maintaining gender inequalities, including unequal patterns of distribution. Only through addressing gendered patterns of behavior and engagement can we effectively address the distribution problem.

IV. WATER

Reliable and relatively easy access to clean water is an issue for billions of people, globally. According to water.org, roughly one of every nine people worldwide (750 million people) does not have access to safe water.[66] Moreover, diarrhea caused by inadequate sanitation, unsafe water and poor hand hygiene leads to the death of 842,000 people annually (2,300 people per day).[67] Not only is water an essential element for survival, health and nutrition, it is also essential for food security. As one United Nations document puts it "water is food," underscoring the point that without adequate water, we cannot grow food.[68]

Women and girl children bear a disproportional burden of the consequences of difficult to access water and unsanitary water sources. In many places, women and girls have the primary responsibility for fetching water; fresh sources of water may be very distant. A study done in twenty-five sub-Saharan African countries between 2006 and 2009 revealed that it is the woman's task to collect water in 62 percent of those households that lack water on their premises. Girl children bore the responsibility in an additional 9 percent of such households.[69] Water.org, drawing on World Health Organization data, reports that water collection is the responsibility of women and children in 76 percent of households in forty-five developing countries. Time spent collecting water is time that is not available to do paid labor, provide caregiving activities, or attend school.[70]

Such fetching is time consuming, and hence presents an obstacle for women and girls in enjoying or performing other activities, including getting an education. In Uganda, for example, women spend a yearly average of 660 hours fetching water. This is equivalent to two months of labor.[71] But even more, a great amount of energy is expended in just this daily task. For example, the United Nations reports that the energy Kenyan women use in the collection of water can be equivalent to 85 percent of their daily caloric intake.[72] In addition, the burden of carrying heavy water pots impacts the health of women and girls; it is the primary cause of pelvic distortion, which can cause death during childbirth.[73]

Access to clean, sanitary water is now seen as a fundamental human right.[74] The United Nations estimates 2.5 billion people lack access to basic sanitation; 1.25 billion of those persons are women.[75] Lack of access to basic sanitation, including clean water and adequate toilets, has significant impacts on women and girls across a range of activities. Girls' school attendance drops without access to adequate and private toilets, especially during menstruation. In many places menstruation is taboo and considered dirty, without adequate access to sanitary, safe, and private conditions to attend to their physical needs during menstruation, many girls and women stay at home and miss school or work. This has negative consequences for the girls and women themselves, in that it impacts their health, safety, education and employment. It also has negative consequences for the economy, since half of the workforce is excluded when women's sanitation needs are not met.[76] Moreover, a lack of access to clean, safe water for drinking and sanitation exposes women and girls to risk of violent assault and rape. Where women have to travel long distances to secure water resources they are vulnerable to attack. And, where women and girls have to seek out isolated spaces to relieve themselves they are vulnerable to physical attack and sexual assault.[77]

Securing access to water is also a critical element of achieving food security. Without water, there is no food. Moreover as noted earlier, women are increasingly responsible for food production in their communities, and reliable access to water is necessary for such production. However, women are often excluded from decision-making processes about water management. Including women in decision-making and political processes is shown to increase the "sustainability of water supplies."[78] For example, if the design of an irrigation system does not take into account the fact that women use the system, it is likely that women will lose access to the land they farm or access to the products of their labor. According to studies done in Cameroon, Laos, Nepal and the Gambia, if women are involved in the design and implementation of the project, it is more likely to be sustainable and effective.[79] A further study done by the International Water and Sanitation Centre (IRC) confirms these results. The IRC examined sanitation and community water projects in eighty-eight communities across fifteen countries. Their findings indicate that those projects in which women were fully involved in both the design and implementation were more successful and sustainable than projects in which women were not fully involved.[80]

In order to achieve the life-sustaining goal of access to clean and safe water, as well as securing sufficient water to produce adequate food, the United Nations and a variety of NGOs advocate for the mainstreaming of gender in water programs as one prong of a complete strategy. What this entails is considering whether there are different implications for women, men, boys and girls, and, if so, how those implications play out with regard to the project, program, or policy.[81] Taking into account the gendered patterns of social behavior and gendered needs of the various individuals within a community is necessary for developing effective solutions. The FAO created the "Passport to Gender Mainstreaming in Water Programmes" as a tool for those engaged in various agricultural water management projects. The passport highlights the fact that where such projects fail to take into account women's experience, knowledge, and needs, it is more likely that women will lose their existing land access and access to products of their labor.[82] The aim of gender mainstreaming is to ensure equal benefits result from the proposed intervention or project development. But, this is only one part of a sustainable solution. The empowerment of women, including through directly addressing unequal gender norms, is an essential second prong of any development project.

Women need to be directly involved in decision-making processes for community development, including importantly water management. Where women are engaged as primary stakeholders in community resources and allocation, their standing in the community is elevated. Moreover, as in the case of food security, removing obstacles (both legal and cultural) to women's land ownership is crucial for sustainable water programs. Whereas gender inequality and resource inequality are mutually reinforcing, so are gender equality and resource equality. Creating access to clean safe water in a sustainable fashion must involve women, with their full participation. Moreover, access to clean safe water is necessary for elevating the status of women and girl-children. Clean water that is easily accessible is a necessary condition for enabling women and girls to get an education and to generate income.[83]

V. CONCLUSION: EMPOWERING WOMEN AND GENDER EQUALITY AS NECESSARY FOR JUSTICE

Gender equality is not only a principal aspect of justice in its own right, it is also a central instrumental good for achieving other fundamental rights such as food security and access to clean, safe water, as we have seen. Thus, the United Nations' emphasis on promoting gender equality and empowering women as its third Millennium Development Goal emphasizes the centrality of gender equality and justice to all forms of equality and moving us closer to a just world.

In order to empower women and girls, and move us closer to a gender just world, the United Nations articulates the following target goals as part of the Millennium Development Goal:[84]

- Increase girls' participation in education, particularly secondary education and beyond

- Increase women's political participation and inclusion in decision-making procedures within communities
- Address the various forms of discrimination women face through law, in employment, and in a sex-segregated labor force
- Reduce violence against women in all its forms
- Address poverty, including extreme poverty, which disproportionately impacts women and girls.

This list will sound familiar to feminists and those concerned with gender inequality as essential goals in securing gender justice, particularly those of us in the Western world. However, as we have seen they are also central elements of securing food security not just for women but also for all people. The feminist goal of achieving a more gender just world includes empowering women and eliminating the traditional barriers to women's full participation in education, employment, politics and civil society; doing these things will lead to better food security for us all.

What is hopefully clear is that securing food justice, for all, especially women, requires much more that simply looking at patterns of distribution, and even more than simply creating more efficient patterns of distribution. Justice requires equal, effective access over the course of women and girl-children's lives. Creating the conditions for such equal access means addressing the patterns of gender inequality that devalue and subordinate women and girl-children. If we aim to solve the problems of unequal gender distribution by simply giving more resources to women (and girls), we won't create a sustainable or fully just situation. As we saw with the microfinance strategy, while it does have some success and has improved the lives of many women, where gender inequality is deeply entrenched, the resources may not actually go to women, or women may not have effective control over the resources, or women may have to take up extra work to make loan payments. Of course, we should keep inventing new strategies to get resources in to the hands of the people who need them most, and microloan programs are a helpful strategy. However, full justice requires addressing the underlying patterns of gender inequality; norms, laws and customs that reinforce the subordination of women must be directly addressed. Until women are men's full social and legal equals, we will continue to see unjust patterns of food security. Food justice, therefore, depends on gender justice.

NOTES

1. This title is a play on a very well-known feminist book, *Fat is a Feminist Issue*, by Susan Orbach, originally published in 1978. A paperback reprint was published in 2006 by Arrow Books.

2. For example, *A Companion to Feminist Philosophy*, ed. Alison M. Jaggar and Iris Marion Young (Malden, MA: Blackwell Publishers Inc., 1998), an edited volume of collected essays, comprising over 600 pages and including fifty-eight distinct entries, does not have a single entry on food or food ethics and gender. A survey of the most common texts for teaching feminist thought, in the West at least, reveals similar results. Insofar as Western feminist texts do mention food, it is often in connection with body-image issues, anorexia and bulimia.

3. See, for example, Martha Nussbaum, *Women and Human Development: The Capabilities Approach* (Cambridge: Cambridge University Press, 2000) and *Creating Capabilities: The Human Development Approach* (Cambridge, MA: Harvard University Press, 2011).

4. As one primary example, see La Via Campesina's website, accessed March 8, 2015, http://viacampesina.org/en/.

5. For a full discussion of distributive justice and the various philosophical debates within the literature, see Julian Lamont and Christi Favor, "Distributive Justice," *Stanford Encyclopedia of Philosophy*, 2013, accessed March 8, 2015, http://plato.stanford.edu/entries/justice-distributive/. For a discussion of international distributive justice, see Michael Blake and Patrick Taylor Smith, "International Distributive Justice," *Stanford Encyclopedia of Philosophy*, 2013, accessed March 8, 2015, http://plato.stanford.edu/entries/international-justice/.

6. See John Rawls, *Justice as Fairness: A Restatement*, ed. Erin Kelly (Cambridge, MA: Harvard University Press, 2001). See also Samuel Freeman, *Rawls* (New York: Routledge, 2007), esp. ch. 3.

7. See, for example, Iris Marion Young, *Justice and the Politics of Difference* (Princeton, Princeton University Press, 1990).

8. See, for example, Nancy Fraser and Axel Honneth, *Redistribution or Recognition: A Political Philosophical Exchange* (New York: Verso, 2003) and Nancy Fraser, *Justice Interruptus: Critical Reflections on the "Postsocialist" Condition* (New York: Routledge, 1997). See also Elizabeth Anderson, "What is the Point of Equality?" *Ethics* 109 (1999): 287–337; Nussbaum, *Women and Human Development*; and Young, *Justice and the Politics of Difference.*

9. Young, *Justice and the Politics of Difference*, 37.

10. Ibid., 38.

11 Fraser and Honneth, *Redistribution or Recognition*, 12.

12. Nussbaum, *Creating Capabilities*, 1. For Sen's work, see Amartya Sen, *Poverty and Famines: An Essay on Entitlement and Deprivation* (New York: Oxford University Press, 1983).

13 Anderson, "What is the Point of Equality?" 312.

14. Ibid.

15. Ibid.

16. The World Health Organization highlights this point on their website. Accessed March 8, 2015, http://www.who.int/trade/glossary/story028/en/.

17. See Peter Singer, "Famine, Affluence and Morality," *Philosophy and Public Affairs* 1 (1972): 229–43 and *The Life You Can Save* (New York: Random House, 2010).

18. The United Nations' Millennium Development Goals are stated here: http://www.undp.org/content/undp/en/home/mdgoverview.html. Accessed April 14, 2015.

19. Rosemary Rop, "Mainstreaming Gender in Water and Sanitation: Gender in Water and Sanitation," World Bank Water and Sanitation Program working paper (2010), accessed March 8, 2015, http://www.wsp.org/sites/wsp.org/files/publications/WSP-gender-water-sanitation.pdf.

20. Ibid.

21. World Health Organization, accessed March 8, 2015, http://www.who.int/trade/glossary/story028/en/.

22. "Gender Equality and Food Security: Women's Empowerment as a Tool against Hunger," Asian Development Bank, 2013, 5, accessed March 8, 2015, http://www.adb.org/publications/gender-equality-and-food-security-womens-empowerment-tool-against-hunger.

23. Ibid, 1.

24. A variety of UN reports cited in this chapter report this statistic. Project Concern International, an NGO focused on disease prevention and sustainable development, claims that women and girls account for 70 percent of the 1.3 billion people living in extreme poverty. Accessed March 8, 2015. See http://www.pciglobal.org/womens-empowerment-poverty/.

25. See Caren Grown, "Missing Women: Gender and The Extreme Poverty Debate" (USAID, 2014), accessed March 8, 2015, http://usaidlearninglab.org/sites/default/files/resource/files/Gender%20&%20Extreme%20Poverty_Missing%20Women.pdf; see also United Nations, "The Millenium Development Goals Report: Gender Chart," (2012), accessed March 8, 2015, http://www.unwomen.org/~/media/headquarters/attachments/sections/library/publications/2012/12/mdg-gender-web%20pdf.pdf.

26. "Gender Equality and Food Security," 1.

27 Ibid., 13.

28 United Nations Development Programme, "Gender and Poverty Reduction," accessed March 8, 2015, http://www.undp.org/content/undp/en/home/ourwork/povertyreduction/focus_areas/focus_gender_and_poverty.html.

29 Elizabeth Ransom and Leslie K. Elder, "Nutrition of Women and Adolescent Girls: Why It Matters," Population Reference Bureau, accessed March 8, 2015, http://www.prb.org/Publications/Articles/2003/NutritionofWomenandAdolescentGirlsWhyItMatters.aspx.

30 Ibid.

31 Ibid.

32. Ibid.

33. "Gender Equality and Food Security," 1.

34. FAO, "FAO in Emergencies Guidance Note: Striving for Gender Equality in Emergencies" (Rome: Food and Agricultural Organization of the United Nations, 2013), accessed March 8, 2015, http://www.fao.org/fileadmin/user_upload/emergencies/docs/Guidance%20Note%20Gender.pdf.

35. Ibid.

36. Ibid.

37. Ibid.

38. Ibid.

39. Addressing the inequalities between men and women with respect to care-giving roles and the devaluing of "women's work" requires institutional change as well as norm change:

> The gender roles associated with the existing division of labor are clearly difficult to challenge without rethinking the broader issues associated with employment patterns and questioning the pressures currently imposed on both men and women. Yet, a number of measures that would alleviate burdens on families would benefit both women and men—e.g., better public transport, water, and energy services, as well as child-care services and institutional care for the sick and elderly. ("Gender Equality and Food Security," 6)

40. Ibid., 1.

41 Alice Aureli and Claudine Brelet, "Water and Ethics, Women and Water: An Ethical Issue," United Nations Educational, Scientific and Cultural Organization, 2004, prepared by Water Aid, the Water Supply and Sanitation Collaboration Council, and Domestos, accessed March 8, 2015, http://unesdoc.unesco.org/images/0013/001363/136357e.pdf.

42. Ibid.

43. Bina Agarwal, "Food Crises and Gender Inequality," DESA Working Paper 107, June 2011, accessed March 8, 2015, http://www.un.org/esa/desa/papers/2011/wp107_2011.pdf.

44. Ibid.

45. Ibid.

46. "Gender Equality and Food Security," 19–20.

47. Ibid.

48. Ibid.

49. Ibid.

50. Agarwal, "Food Crises," 11.

51. "Gender Equality and Food Security," 24, Figure 7.

52. "Gender Equality and Food Security."

53. Ibid.

54. Ibid.

55. Ibid.

56. Ibid., 28.

57. Agarwal, "Food Crises," 12.

58. Ibid.

59 FAO, "FAO in Emergencies Guidance Note."

60. Agarwal, "Food Crises," 15–16.

61. See http://www.cgap.org/about/faq, accessed March 8, 2015.

62. "Gender Equality and Food Security," 33.

63. Ibid.

64. Ibid., 34.

65. Ibid.

66. See http://water.org/water-crisis/water-facts/water/, accessed March 8, 2015.

67. Ibid.

68. UN Water, "Water is Food," accessed March 8, 2015, http://www.unwater.org/worldwaterday/learn/en/?section=c325501.

69 See UN, "Gender Chart."

70. http://water.org/water-crisis/water-facts/women/, accessed March 8, 2015.

71. Domestos, Water Aid, and WSSCC, "Why We Can't Wait: A Report on Sanitation and Hygiene for Women and Girls," accessed March 8, 2015, http://www.zaragoza.es/contenidos/medioambiente/onu/1325-eng_We_cant_wait_sanitation_and_hygiene_for_women_and%20girls.pdf.

72 United Nations Development Programme, "Gender and Poverty Reduction."

73. Aureli and Brelet, "Water and Ethics."

74 Domestos, Water Aid, and WSSCC, "Why We Can't Wait."

75. Ibid.

76. Ibid.

77. Ibid.

78. FAO, "Passport to Mainstreaming Gender in Water Programmes: Key Questions for interventions in the Agricultural Sector," Food and Agricultural Organization of the United Nations, 2012, accessed March 8, 2015, http://www.fao.org/docrep/017/i3173e/i3173e.pdf.

79. Ibid.

80. UN Water, "Gender, Water and Sanitation: A Policy Brief," accessed March 8, 2015, http://www.un.org/waterforlifedecade/pdf/un_water_policy_brief_2_gender.pdf.

81. FAO, "Passport to Mainstreaming Gender."

82. Ibid.

83. UN Water, "Gender, Water and Sanitation: A Policy Brief."

84. United Nations, "Goal 3: Promote Gender Equality and Empower Women," accessed March 10, 2015, http://www.un.org/millenniumgoals/gender.shtml.

Chapter 9

Meat Eating and Masculinity

A Foucauldian Analysis

Nancy M. Williams

This chapter presents a Foucauldian analysis of the traditional association between masculinity and meat eating. In the United States and in most Western nations, consumers are disciplined to eat meat on a daily basis. While both men and women consume meat, the cultural link between masculinity and meat eating makes men especially vulnerable to subjugating themselves to self-surveillance and the crushing effects of normalization. I will argue that the conceptualization of meat eating as a masculine exercise is itself a mechanism of docility and the failure to critique and resist this practice may put men's bodies at significant risk for ill health as well as curtailing self-management and agential possibilities.

In order to lay the groundwork for my analysis I reference Sandra Bartky and Cressida Heyes and their appropriation of Michel Foucault's theory about the body and how, on their view, it provides an effective tool for deconstructing the disciplinary techniques associated with the construction of feminine identity and embodiment, including the disciplinary practice of dieting. Like women, men confront societal pressures regarding their alimentary practices. They are encouraged to eat meat, especially red meat. However, as I will show, scientific evidence suggests that a diet high in meat may contribute to poor health, including heart disease. I will also show that the normalization effects in this case jeopardize men's agential possibilities or their ability to make genuine and meaningful choices about food and self-identity. Meat eating does not stand as authentic masculine power as the meat industry so often advertises; instead it is a disciplinary practice pushed by agribusiness (and others) that produces subjected bodies and identities. As such, it may seem that traditional masculine identity in this context is profoundly unfree. However, given Foucault's complex account of power and the possibility of resistance and subversion at least on the microlevel, I argue that this is not necessarily the case. On my view, the epistemological tensions associated with meat eating can be the impetus where intrasubjective tension arises and where individuals can begin to question dominant discourse and forge a more independent path. Choosing a vegetarian/vegan diet may be understood therefore as an embodied protest against hegemonic institutional practices. While this is certainly the case for women who choose a meat-free diet it is especially so for men.

I. FOUCAULT: MODERN POWER AND SELF-POLICING

In *Discipline and Punishment*, Foucault lays out his account of modern power and its effect on subjects.[1] We have moved from sovereign power as the prevailing method of exerting power onto others in the mid-eighteenth century to a disciplinary power that is no longer situated within any one Hobbesian ruler or institution. His account challenges the notion that power merely comes from the "outside," from up above in some top-down, unilateral fashion yielding its negative, repressive force onto subjects by way of established laws and mechanisms of censorship. Instead, in the modern age, power comes from everywhere, as it "circulates through progressively finer channels, gaining access to individuals themselves, to their bodies, their gestures and all their daily actions."[2] To give a clearer sense of the nature of power and its effects on subjects, Foucault appeals to philosopher Jeremy Bentham's design for the model prison, the Panopticon. The plans for this prison call for a circular structure and at its center lies a tower with wide windows that opens to the inner side of the structure. The periphery is made up of individual cells, each with a window facing the center tower and facing the outside which allows backlight in the cell making the prisoner visible within the cell. "All that is needed then," Foucault writes, "is to place a supervisor in a central tower and to shut up in each cell a madman, a patient, a condemned man, worker or schoolboy."[3] Although the prisoner is separated from the other inmates (to prevent rebellion) he may be under constant surveillance from the center tower and the effect of this is "to induce in the inmate a state of conscious and permanent visibility that assures the automatic functioning of power."[4] The prospect of being surveyed by the "tower's" disciplinary gaze constructs a heightened self-consciousness in the inmate. He becomes acutely aware of his own behavior as he moves about in an environment of perpetual surveillance. His "visibility is a trap."[5] Since he does not know whether he is being watched and judged to be acting in compliance, he polices himself and thus "becomes the principle of [his] own subjection."[6] The tight, disciplinary control of the jailer (his gestures and habits) lies therefore within himself. It is the fact of being constantly seen, "of being able always to be seen," Foucault writes, "that maintains the disciplined individual in his subjection."[7] Possible surveillance from another ensures self-surveillance and disciplined subjects with the overarching goal of normalization (i.e., the process of homogenizing the population). For Foucault the Panopticon's architecture represents an effective and efficient form of controlling a large population because the disciplinary gaze fosters subjects who are attached, unwittingly perhaps, to their own subjection. No need for a supervisor in the central tower when subjects act as their own disciplinarians or jailers.

While we may find this mode of control in established institutions such as hospitals, schools, factories and the military, Foucault's main point is that we are all by virtue of entering an observed space subject to this form of disciplinary power. In addition to abiding by the law of the land, one is expected to conform to certain social norms about how we should present our embodied selves to the world. Although historians have written about the body within the context of pathology, metabolisms, physiological processes and even a site of germs and viruses, Foucault urges us to think about how "the body is also directly involved in a political field; power relations have an immediate hold upon it; they invest it, mark it, train it, torture it, force it to carry our tasks, to perform ceremonies, to emit signs."[8] Cultural practices always already inscribe "on bodies and their materiality, their forces, energies, sensations, and pleasures."[9] It is a politicized medium, a locus of subjectivity and as I will show later a site of possible resistance. Foucault recounts how in the eighteenth century the individual body was discovered as "object and target of power," one that could be "manipulated, shaped, trained, [and] which obeys, responds, becomes skillful and increases its forces."[10] The manipulated body becomes the docile body, one that may be "subjected, used, transformed and improved."[11] By virtue of the public gaze and self-surveillance the smallest detail of the body, its movement, appearance and functioning, is monitored. These meticulous operations of the body can be called "disciplines," those behaviors which regulate movements and produce "subjected and practiced bodies, 'docile' bodies."[12] Under the tower's gaze the conquered body or one's regulated and monitored behaviors begin to seep into the subject's sense of self. Contrary to the traditional Cartesian formulation where the dualist posits the body as separate from the mind, Foucault's concept of the body is not conceptually incidental to one's sense of self or personal identity, instead the body is constitutive of it. Our bodies exist and navigate through the social world by way of various habits and practices to such a degree that those practices begin to construct our self-identities. In other words, disciplinary techniques target both the body and mind. To push back against these disciplinary forces or to resist the project of docility is not without negative effects however. Resistance can bring about an "infra-penalty," a form of disciplinary punishment that aims to correct behavior by relying on our fear of being humiliated, excluded and labeled as different, deviant or silly.[13] This form of disciplinary punishment, on Foucault's view, aims to construct proper identities (e.g., a "normal" person with "normal" sexual desires for instance). The fear of not being perceived as "normal" may instill a general paranoia in the subject where she constantly monitors her actions and desires against a perceived or assumed intangible social norm. Over the long term, this paranoia can make us all too dependent on experts or authorities to inform us about how we should behave. Other disciplinarians, including teachers, parents, friends, coworkers, spouses, lovers, family, mass media and corporate interest all have the potential to unduly limit self-governance and self-creation by virtue of this form of penalty. It is in this way that modern power has a strong grip on individuals.

II. FEMINIST APPROPRIATIONS

Some feminist scholars find Foucault's account to be an effective theoretical tool to critically examine the modern disciplinary technologies that help construct feminine embodiment.[14] On their view, his notion of the body, as it is affected by social norms and regulated by self-policing, is particularly well suited to the ever-persistent issue of female body image. In her classic essay, "Foucault, Femininity and the Modernization of Patriarchal Power," Sandra Bartky appropriates Foucault's theory to examine the prescribed feminine norms and the implications it has on women's minds and bodies.[15] She does so only after exposing Foucault's gender blindness. In his analysis about modern power Foucault fails to make clear distinctions between male and female bodies or the disciplinary practices that sculpt feminine and masculine disciplinary practices, and while "women, like men, are subject to many of the same disciplinary practices Foucault describes," Barkty points out that "he is blind to those disciplines that produce a modality of embodiment that is peculiarly feminine."[16] Foucault's model essentially speaks to an ambiguous body, one not marked or pressed upon by gender. "We are born male or female," Bartky writes, "but not feminine or masculine. Femininity [and masculinity] is an achievement, a mode of enacting and reenacting received gender norms which surface as so many styles of the flesh."[17] In other words, Foucault's account of disciplinary practices and docile bodies may yield greater insight into the phenomenology of subjectivity and power if we also keep in mind the politics of gender. To proceed otherwise, to assume that female and male bodies are affected in exact ways, is to overlook the disciplinary nature of gendered discourse and expectations.

Bartky attempts to correct Foucault's androcentrism and identifies three specific categories of disciplinary practices that constitute feminine bodies:

(1) practices that aim to produce a body of a certain size and shape, (2) practices that elicit a specific repertoire of gestures, and (3) practices that present the feminine body as an ornamental surface.[18] The first category illustrates how the feminine body type is culturally prescribed and almost impossible for most women to achieve without participating in a variety of disciplines, including dieting, exercise and cosmetic surgery. Of diets, Bartky writes, "Dieting disciplines the body's hungers: Appetite must be monitored at all times and governed by an iron will. Since the innocent need of the organism for food will not be denied, the body becomes one's enemy, an alien being bent on thwarting the disciplinary project."[19] The female body becomes a thing to manipulate, shape, train, work on, subjugate and transform from its original imperfect state, and eventually a political site upon which the effects of feminine normalization must be made apparent. In some cases losing weight can be a matter of serious health consequences since obesity has been linked to certain diseases and chronic ailments; however, the worry here is about the ideal dieter who, feeling fat, documents every calorie, obsesses about her weight and defines her identity entirely by the dictates of the "tyranny of slenderness."[20] Since the standards of what counts as the ideal female body are almost impossible to achieve for most women, she may live much of her life with the overwhelming feeling that her body is permanently deficient no matter the amount of effort or expense. She may also think of herself as less than, defective, and unworthy. If the dieter persists, without pause and critical reflection, her compulsion may transpire into harmful consequences, including eating disorders like bulimia and anorexia nervosa, sleeping problems, malnutrition, atherosclerosis and high-risk cosmetic surgery. As Foucault made clear in his discussion about the effects of the disciplinary gaze, the distinctive feature of technologies is the way in which they operate insidiously on both mind and body. Although the pursuit for weight loss may begin from the "outside" through the disciplinary gaze of the public eye, part of the effectiveness of these practices relies on its moving "inside" through self-monitoring. The dieter cultivates a heightened self-consciousness about her body's appearance as if she is always being watched and assessed by others. She, like the inmate in the Panopticon prison, internalizes the tower's gaze and as such "the automatic functioning of power is assured."[21] When this kind of self-management reduces one's ability to imagine other ways of relating to oneself (and others) it may thwart the possibility of genuine and meaningful choices. Her effective practices of critique (and autonomy) are under threat and the possibility of creative self-transformation (i.e., accepting alternative body shapes and sizes) truncated. In short, her docile body gives way to a subjugated identity, one bound by the dictates of the disciplinary gaze and obedience to patriarchy.

The limited mobility and spatiality of women's gestures are discussed in the second category where women are expected to move their bodies in a confined manner and take up less space when walking and sitting. Bartky, in the third category, describes the countless beauty regiments that many women engage in: hair removal, buffing, waxing, electrolysis, applying an endless supply of skin-care products (e.g., masks, astringents, lotions and creams), and perfecting the skill of proper make-up application. In general, women's bodies should be thin and never show signs of aging. It is, according to Bartky, "a body of early adolescence."[22] These practices, on her account, ultimately produce an inferiorized body, one that "speaks eloquently, though silently, of [women's] subordinate status in a hierarchy of gender."[23] Bartky successfully demonstrates, I think, the ways in which Foucauldian disciplinary practices aim to construct feminine embodiment and how the central technique of disciplinary power, namely self-policing, explains women's collusion with patriarchal standards of femininity. We are our own disciplinarians to varying degrees, succumbing to the forces that continue to define feminine embodiment for us rather than by us. The pressure to achieve the ideal feminine body size and shape can be overwhelming given the possible societal sanctions (e.g., fat phobia and shaming). The proclaimed rewards of being thin are also considerable; being thin, we are told, may enhance our job opportunities and dating prospects.[24] As Susan Bordo explains,

> Preoccupation with fat, diet, and slenderness are not abnormal. Indeed, such preoccupation may function as one of the most powerful normalizing mechanisms of our century, insuring the production of self-monitoring and self-disciplining "docile bodies" sensitive to any departure from social norms and habituated to self-improvement and self-transformation in the service of those norms.[25]

So, it is understandable why weight loss programs like Weight Watchers continue to enjoy great financial success with a membership comprising mostly of women.[26]

Cressida Heyes examines the highly profitable weight loss industry, particularly Weight Watchers, through a Foucauldian lens in order to shed light on the complex way dieting regiments can both imprison and liberate women.[27] The tension between the acquisition of new dietary knowledge and skills and the overwhelming effects of (feminine) embodied normalization is one in which Heyes believes we ought to investigate more closely because it may be the locus of embodied resistance and self-transformation. This is an important observation because it offers a strong rebuttal to a major worry among some of Foucault's critics. A common criticism leveled against Foucault is that his theory fails to offer an adequate explanation as to how self-transformation and agency is possible. If the individual represents the site upon which modern disciplinary power produces complicity then Foucault's body is essentially passive with limited possibility for self-regulation. If it is the case that individuals can never fully escape the effects of normalization, how is resistance or agency possible? How is it theoretically possible for women to break free from the effects of embodied normalization when we are always caught up in the ever-tightening web of gender norms? Foucault, in his later work, *The History of Sexuality Volume 1*, frames these worries as a misunderstanding of what he means by power:

> Should it be said that one is always "inside" power, there is no "escaping" it, there is no absolute outside where it is concerned, because one is subject to the law in any case? Or that history being the ruse of reason, power is the ruse of history, always emerging the winner? This would be to *misunderstand* the strictly relational character of power relationships.[28]

He explains that the effect of normalization on the subject is not a one-way relation where the former has a complete hold over the latter. He asks us to have a more complex understanding of power, one that is relational (i.e., produced among persons, institutions, things and groups of persons), unstable, mobile and shifting. Power comes from below, up above and within immanent relations: "it is produced from one moment to the next, at every point, or rather in every relation from one point to another."[29] The nature of power, therefore, is one that vacillates among its constituents. In addition, power is not inherently bad or good, it can be positive and productive as well as an apparatus for domination.[30] He argues that we need to stop thinking in terms of power as a force that is always negative, as something that necessarily censors, represses, or conceals. While it can define or constrict our somatic selves, power can also be the mechanism for promoting increased options and greater freedom among individuals. Power, according to Foucauldian scholar Margaret McLaren, "functions ambivalently for Foucault. In its negative aspect it serves to limit, to dominate, to normalize: this traditional understanding of power is akin to what feminists call 'power over.' In its productive, positive aspect, power creates new possibilities, produces new things, ideas and relations; this is akin to what feminists call 'empowerment.'"[31] "Where there is power there is resistance," writes Foucault, "[and] these points of resistance are present everywhere in the power network."[32] To be clear, he does not advocate for the possibility of complete immunity from the effects of normalization or a social criticism originating from some transcendental (epistemological or moral) basis, but rather a localized (or microlevel) struggle. Consider for instance Heyes's examples of acts that subvert embodied norms: body modification (e.g., scarification, branding, flesh hanging, or subdermal implants), yoga (as a different way to understand one's relationship to the body than conventional dieting regimes), and male ballet dancers. Even Bartky points to radical lesbian communities and female bodybuilders as examples of what women can do to counter the hegemonic patriarchal forces pressed against us.

Borrowing from this Foucauldian conception of power, Heyes argues that Weight Watchers's success can be explained, in part, by the new capacities and skills members acquire. Some members, Heyes reports, would utilize useful tips and dietary strategies from the program as a way to cultivate a "space" for playfulness and spontaneity about their food choices. "The real women I met," Heyes writes, "were often aware that they could learn from Weight Watchers *without becoming the projected unified subject of its regime*. Central to this awareness is the possibility of uncoupling new capacities from docility (e.g., documenting and observing strategies) and recruiting these capacities to care of the self."[33] The disciplinary practices of weight management were taken to improve one's awareness of one's own habits and feelings as well as taking on a new sense of responsibility for choices about how to live well. While the calorie-documenting and weight-watching techniques may demonstrate the constrictive nature of power over our somatic selves, the improved awareness and sense of responsibility derived from one's dietary surveillance could be understood as the productive and positive aspect of power. On the one hand, these new capacities may deepen its members' dependence on the organization and forestall the possibility of agency. On the other hand, this form of self-discipline may assist others to engage in techniques of self-care and make room for attitudes that are less confining and that offer a more positive understanding of the relation to oneself. So, on Heyes's view, resistance resides in the complex vacillating relational character of power. The place where self-discipline and new capacities meet is the "fissure feminists should exploit."[34]

Heyes's account as to how one can resist complete normalization parallels nicely with Foucault's nuanced and sophisticated understanding of power; however, there may be another fissure we might want to exploit, one that could give rise to critique and resistance: when the disciplines themselves are fraught with epistemological tensions or contradictions. Because disciplinary practices are often slippery and fluid, (ideological, moral and epistemological) fissures sometimes appear and opportunities for critique and embodied resistance emerge. When norms are characterized by inconsistencies, ambiguity and/or contradictions, tears in "the fabric of our epistemological web" appear and "the practice of critique emerges."[35] And critique, questioning status quo practices, is at the heart of Foucault's theory with respect to the possibility of resistance and insubordination. For Foucault,

> Critique is the movement by which the subject gives himself the right to interrogate truth on its effects of power and question power on its discourses of truth. . . . Critique will be the art of voluntary insubordination, that of reflective

indocility. Critique would essentially ensure the desubjugation of the subject in the game of what we could call, in a word, the politics of truth.[36]

In other words, critique makes way for resistance and agential possibilities. When one's identity is tied up by the norms of gendered embodiment (and specific alimentary practices) and those norms are characterized by inconsistencies intrasubjective tension may arise in the individual and as such greater critical awareness and freedom emerges. My point is the recognition of competing normative claims may loosen the grip the dominant discourse has on individuals in much the same way that new capacities can encourage individuals to develop creative techniques of self-care.

With respect to dieters, agential possibilities arise, for example, when individuals grapple with the epistemological tensions associated with weight management regimes. One basic apparatus carrying the success of the weight loss industry is the height-weight chart. The idea that there is an objective standardized range which each individual's weight must fall in order to be considered healthy is "bogus," according to Heyes, because "standardized weight tables are artifacts of actuarial insurance company definitions that were themselves never based on comprehensive statistical information. They have changed over the years for no medical reason, and have become a better measurement of social acceptability than morbidity or mortality."[37] There is also an assumption that certain health problems will automatically disappear upon weight loss. While obesity has been associated with certain health issues, including heart disease and certain cancers, to suggest that the "weight loss itself is a stand-in for health" neglects the other factors that play a role in good health.[38] According to Glenn Gaesser, author of *Big Fat Lies*, "No study yet has convincingly shown that weight is an independent cause of health problems."[39] Gaesser goes on to show that yo-yo dieters, "who make up about 90 percent of the dieters in this country . . . have a risk for type II diabetes (the most common kind) and for cardiovascular disease that is up to twice that of 'overweight' people who remain fat."[40] In other words, yo-yo dieting is probably more dangerous than carrying those extra pounds. Furthermore, many dieters are yo-yo dieters because most (about 95 percent) of all diets fail over the years immediately following weight loss (and Weight Watchers is well aware of this failure rate given their successful re-enrollment programs). These epistemological fissures can be the locus where intrasubjective tension arises and where individuals begin to question the dominant discourse and craft instead a more independent attitude toward one's somatic self. So, in addition to the possibility of uncoupling self-discipline practices from docility as Heyes notes in her analysis, I suggest that the inconsistencies and contradictions associated with norms can also act as the locus for agency and self-transformation. They offer individuals the opportunity to resist the complete crushing effects of dietary normalization. I am not claiming that critical reflection is sufficient for self-transformation and freedom only that it is a first step toward a way of being in the world that mitigates the normalizing effects of disciplinary power.

With Bartky and Heyes, we have a glimpse into the ways in which women's bodies including their diets constitute their feminine identities. We saw how dietary practices form what Foucault calls disciplinary power and the possible deleterious effects those practices may have on the female body and mind. Of course, the pressure to conform to gender norms, including alimentary ones, is not unique to women. Men too are subject to alimentary norms: they are encouraged to eat meat. In "Masculinity as Homophobia," Michael Kimmel, a leading scholar on masculinity, argues that American men are socialized to conform to strict and narrow definitions of manhood and that the pressure to be "man enough" constitutes a fear of other men.[41] Men fear that other men will think of them as too feminine if they do not abide by traditional notions of masculinity and this fear ultimately perpetuates homophobia and limits self-expression. This fear, according to Kimmel, "dominates the cultural definitions of manhood" and it starts at adolescence when "[boys] learn that [their] peers are a kind of gender police, constantly threatening to unmask [them] as feminine, as sissies."[42] Young boys learn early that they must keep a constant vigilance on their manly front, "making sure that nothing even remotely feminine might show through."[43] Self-surveillance and the policing of other males, as Kimmel states, is everywhere and involves everything a man does, "What we wear. How we talk. How we walk. What we eat. Every mannerism, every movement contains a coded gender language."[44] The pressure to measure up to traditional notions of masculinity, according to Kimmel, results in a manhood that is "so chronically insecure that it trembles at the idea of lifting the ban on gays in the military, [and] that is so threatened by women in the workplace that women become the targets of sexual harassment."[45] I would also add that the pressure to abide by traditional notions of masculinity results in a manhood that trembles at the idea of consuming less or no meat.

III. MASCULINITY AND MEAT: DOCILE BODIES AND SUBJUGATED IDENTITIES

Although both women and men are disciplined in the West to consume a meat-based diet, men make for the most obvious target of this disciplinary practice given the cultural link between traditional notions of heterosexual masculinity and meat consumption. Real men (i.e., men who exhibit traditional masculine behaviors) eat meat; that is, strong (i.e., healthy), virile, independent men consume animal flesh on a regular basis. Meat (but especially beef) is a consummate male food, and "an exemplar of maleness."[46] Meat is, according to Carol Adams, "a symbol and celebration of male dominance [and] . . . failure of men to eat meat announces that they are not masculine."[47] Recent empirical surveys and sociological studies support Adams's theory. According to the American Meat Institute, of those who consume meat, men on an average eat 6.9 oz. of meat per day and women eat 4.4 oz.[48] Researchers from Bellarmine University in Kentucky found that men believed eating meat makes them more manly and at the University of British Columbia vegetarian men were considered effeminate.[49] It is clear that meat consumption is a central part of the enactment of normative masculinities and widely considered to be essential for masculine bodies and vitality.

Power, Foucault reminds us, entails a systematic web of interlocking mechanisms each feeding off one another and this is certainly the case here. Marion Nestle, professor of food science and public health at New York University and former policy advisor to the Department of Health and Human Services, gives firsthand accounts of how for decades the meat industry strong arms the government's effort to construct objective and helpful dietary guidelines (e.g., ChooseMyPlate.gov).[50] She also exposes the tradition of filling USDA appointments with people who have direct ties to and a vested interest in the meat industry. In addition, farmers who supply meat products receive considerable government subsides, about 200 billion dollars between 1995 and 2010. Farmers who grow fruits, vegetables and tree nuts, on the other hand, receive no regular direct subsidies.[51] In terms of marketing, the link between masculinity and meat eating is, "widely circulated in popular discourse and promoted by those who have vested interests in encouraging meat consumption, such as beef producers."[52] Nestle explains that the beef boards, "design campaigns to boost demand for red meats and meat products; encourage consumers to view beef as wholesome, versatile and lower in cholesterol; and educate doctors, nurses, dietitians, teachers and the media about the nutritional benefits of beef."[53] Consider, for instance, *Men's Health*, a widely circulated magazine in the United States and abroad, that provides lifestyle advice for men on virtually every aspect of living, including sex, fashion and food. The food advice as might be expected tends to coincide with traditional notions of what men should eat.[54] In the September 2000 issue, for instance, the article "138 Things a Man Should Never Apologize For" advises men to never apologize for: "Liking McDonalds, Not offering a vegetarian alternative . . . Laughing at people who eat trail mix."[55] One can also find in the same issue an article entitled, "Your Dinner Personality," where pictures of food are accompanied by brief text explaining the associated personality. The t-bone steak is "something a big man would eat," and a picture of John Wayne is placed next to it. The filet mignon caption reads,

"Classy, likes to indulge," and consuming burgers apparently demonstrates that "the guy can be himself." The pasta primavera, on the other hand, is a "dull choice, dull guy." Vegetables are not only portrayed as dull but as effeminate: "Vegetables are for girls. . . . If your instincts tell you a vegetarian diet isn't manly, you're right."[56] The magazine also plays up, at least implicitly, the prevailing myth that animal-based protein is necessary for optimal health and muscle development with text such as: "Meat has big advantages over all other foods: It packs muscle-building protein," or "Meat is loaded with the protein needed to build new muscle."[57] Celebrity vegan bodybuilders or athletes are noticeably absent, such as the 2009 Mr. Natural Universe, Billy Simmonds, or Patrik Baboumian who broke the world record for the most weight carried, and of course Arian Foster, of the Houston Texans, who in 2010 led the NFL in rushing yards and touchdowns and was awarded "NFL Strongest Performance" that same year.[58] To be fair, in 2012 *Men's Health* published an article detailing the successful careers of vegan male athletes albeit with a disclaimer that "the diet can be dangerous."[59] No disclaimer however was found with the article promoting heavy (red) meat consumption and given recent scientific studies it may be more appropriate to do so. These examples point to the theoretical basis of this project, that is, the male body is directly involved in a political field. These dietary practices inscribe upon his body a political meaning (i.e., what it means to be a real man) and as such encourages the self-policing and normalization of masculine bodies by virtue of alimentary disciplines.

Previously we examined the association between femininity and thinness and the possible deleterious effects this cultural link has on women's bodies and sense of self. So, how might the link between masculinity and meat eating harm bodies and jeopardize identities (or self-management) in men? Consider for instance the ubiquitous claim that meat consumption makes men strong and healthy. As we saw in *Men's Health* the consumption of meat is often justified as masculine because it will build muscular strength. Beef is particularly symbolic since it comes from the largest and most muscular (albeit herbivore) farmed animal. However, by most standards, the promotion of red meat for men cannot be considered a health goal. The Physicians Committee on Responsible Medicine points to studies that suggest "people who avoided meat were much less likely to develop [cancer]."[60] According to Will Courtenay, internationally recognized expert in men's health, "the average man's diet is a major contributor to heart disease and cancer . . . [and] males of all ages consume more saturated fat and dietary cholesterol than females . . . [and] . . . consume less fiber and fruit and fewer vegetables than women . . ."[61] One in three men will develop some major cardiovascular disease before age sixty as opposed to one in ten women.[62] According to the American Cancer Society, "Men who eat a lot of red meat or high-fat dairy products seem to have a greater chance of getting prostate cancer."[63] Although scientists have differing hypotheses as to why men are more prone to develop heart disease (e.g., increased stress, sex-specific hormones) it is reasonable to consider high meat consumption as a possible factor.[64] In a recent Harvard study, for instance, it was revealed that meat contains a molecule, L-carnitine, which can contribute to heart disease.[65] In another study, the positive association between meat consumption and fatal ischemic heart disease was stronger in men than in women and, overall, strongest in young men. For forty-five to sixty-four-year-old men, there was approximately a threefold difference in risk between men who ate meat daily and those who did not eat meat.[66] So, contrary to the prevailing adage that high meat consumption builds healthy masculine bodies, to eat like a man is to subject oneself to chronic illnesses and early mortality.[67] Finally, meat consumption is strongly linked to erectile dysfunction.[68] Fifteen to thirty million American men are affected by erectile dysfunction and yet little attention is paid to the dietary influences that can bring it about.[69] If one suffers from atherosclerosis (i.e., hardening of the arteries), which is linked to dietary cholesterol (i.e., the consumption of animal products), it is reasonable to assume other organs can be affected. In the face of scientific and medical evidence, the societal construction of (male) meat eaters as healthy, strong and virile is in fact illogical.

The fact that a high-meat diet is not a health goal is well known in our society. Most people know that frequent trips to the Burger King drive thru do not promote optimal health, and yet, many continue to perform their prescribed meat-eating role. One possible explanation may be that we are duped by the pervasive advertising claims where "certain social realities are made systematically obscured by an internally coherent ideology whose propagation has material benefits for the dominant group" (e.g., agribusiness).[70] Another possibility points to akrasia, that is, we know burgers and chili fries are bad for us, but we just cannot help ourselves. We simply give in to culinary temptation. While pervasive advertising and a lackluster will play a role here, the consumption of meat (in this gendered context) transcends health concerns. Choosing a menu item sometimes involves more than meeting one's need for nutrition. What we choose says something about who we are or what we hope to be. All of us constitute our identities through what we eat and men who do not eat meat in our culture are often described as effeminate, abnormal, or homosexual; and in a misogynistic and homophobic society it is understandable why many men bow to the pressure and police their dietary practices in such rigid (and harmful) ways. To rebel against meat eating (to adopt a plant-based diet for instance) is to possibly bring about an "infra-penalty," a form of disciplinary punishment that aims to correct our behavior by relying on our fear of being humiliated, excluded and labeled as different. In this way, disciplinary punishment aims to construct proper (male) identities and the long-term effect of this normalization is a docile body (riddled with preventable ailments) and a docile mind (or an identity held captive by limited self-governance). When this kind of self-surveillance reduces men's ability to imagine other ways of relating to themselves and others, it impedes the possibility of crafting genuine and meaningful dietary choices and subjectivities; instead they become strict conformists to prescribed roles. Masculine identity in this context is thus profoundly unfree. Meat eating does not constitute authentic (masculine) power as the industry so often portrays; instead it is a disciplinary practice which represents the power that agribusiness has over us. In short, the underlying paradox of associating meat eating with masculine independence is that it tends to narrow behavioral (dietary) options for men and generate an individual who is a strict conformist to alimentary norms. Consuming meat represents social conformity; it is, in Foucauldian terms, a disciplinary practice pushed by agribusiness (and others) that produces docile bodies and subjected identities, and given the link with masculinity, this is especially so for men. The prevailing message that meat eating represents masculine virility, strength and independence (that is, no one will tell me what I can or cannot eat) is in truth a mechanism of docility and subjugation and it puts men's bodies and agency at risk.

IV. CONCLUSION: AGENTIAL POSSIBILITIES

However, being trapped in this traditional model of masculinity is not a foregone conclusion. Recall in its productive positive aspect power creates new possibilities and produces new things, ideas and relations. From a Foucauldian point of view the individual is always shaped but not necessarily determined by disciplinary power. Power produces not only docile bodies, but resistant bodies. As embodied persons we are both normalized and agential. Moreover, if my theory is correct, the epistemological fissures associated with the meat-eating model can be the locus where intrasubjective tension arises and where individuals can begin to question the dominant discourse and craft a more independent attitude toward their somatic self. While many do not investigate contradictory norms that is not to say it is impossible to do so. The Straight Edge (sXe) Movement, a subculture of young, usually extensively tattooed and pierced, macho men (although women can also be members) who listen to loud punk music and dress in provocative black leather outfits is a case in point. At first blush they exhibit prototypical maleness with aggressive postures and language, but with one notable exception, many espouse a vegan diet. Their philosophy, in general, is to commit to a lifestyle free of drugs, alcohol and the consumption of animal products. Vegetarianism or veganism becomes a logical extension of their overall philosophy to engage in a flourishing and healthy life free from harmful practices. The Straight Edge Movement is one example of how masculinity can be redefined in the face of critical reflection and it also illustrates how dietary choices represent agential possibilities. I witness this kind of self-transformation among some of my male students when, after taking one of my classes on animal ethics and/or the gender politics of food, they adopt a vegetarian/vegan diet. When I inquire as to why they no longer eat meat (despite societal pressure to do so) they tell me that they did not want to support an industry that profited from animal cruelty and environmental degradation; instead they hope to exercise a more consistent ethical life by adopting a meat-free diet. Health reasons propelled others. Some students question the ideological inconsistency in consuming factory-farmed animals as a demonstration of masculine power. What is so manly, they ask, about eating sick, weak animals? Examining some of the contradictions associated with masculinity and meat eating also carved out an intellectual space for students to critically reflect on their prescribed dietary practices and the ways in which agribusiness encourages self-surveillance and docility. While most continued to eat meat, a minority of male students began to see vegetarianism/veganism as an embodied political protest against the politics of truth regarding alimentary norms. They underwent a kind of "undisciplining and redisciplining" of themselves: the rejection of conventional dietary practices represents a form of undisciplining, while veganism, though a form of self-discipline, departs from the kind done in the name of compliance and the production of docile bodies.[71]

The conversion to vegetarianism/veganism for these young men encapsulates the kind of embodied self-transformation and local resistance that Foucault's theory alludes to. The struggle against those who have power over us begins at the microlevel or, in this case, at the dinner table. Choosing a meat-free diet therefore can be understood as an instance of agential possibility within a Foucauldian framework. I am not claiming that Foucault's work endorses vegetarianism/veganism as the embodied or dietary ideal. My point is vegetarianism/veganism might be understood as an embodied protest against hegemonic regulation; it represents a nonconformist identity that is open to novel ways of being in the world. This is certainly applicable with women who choose a meat-free diet, but it is especially so for men given the social context or the Panopticon they inhabit. It is also important to point out that my thesis does not imply that men are therefore made oppressed by this form of disciplinary control. In order to determine whether someone's suffering is an element of what it means to be oppressed we must ascertain the context in which that suffering is taking place. In this case men are not faced with an "enclosing structure of forces and barriers" which diminishes their social standing as men in a male-dominated society.[72] Men as a group continue to benefit significantly from patriarchy both materially and culturally and with regards to access to power.

Examining our alimentary identities and practices through a Foucauldian lens opens up new ways of understanding food. Our food choices possess philosophical significance that is characterized by complex notions of personal identity, power and the possibility of radical self-transformation. It may be difficult to challenge gendered and alimentary norms. By pushing back against the forces of normalization we render ourselves marginalized, different and outcasts. However, the willingness to deconstruct the epistemological fissure between "strong, virile, independent men eat meat" and the medical evidence which suggests the contrary may be an important part of what it means to engage in authentic living. We owe it to ourselves and the selves we may become to reject docility and to break free from certain norms, especially those that perpetuate harmful practices.

NOTES

1. Michel Foucault, *Discipline and Punishment*, trans. Alan Sheridan (New York: Pantheon, 1977).

2. Michel Foucault, "Body/Power and Truth and Power," in *Michel Foucault: Power/Knowledge*, ed. C. Gordon (U.K.: Harvester 1980), 151.

3. Foucault, *Discipline and Punishment*, 200.

4. Ibid., 201.

5. Ibid., 200.

6. Ibid., 203.

7. Ibid., 187.

8. Ibid., 25.

9. Michel Foucault, *The History of Sexuality, Vol. 1: An Introduction*, trans. Robert Hurley (New York: Vintage, 1980), 155.

10. Foucault, *Discipline and Punishment*, 136.

11. Ibid., 137.

12. Ibid., 138. Foucault goes on to explain that "'Discipline' may be identified neither with an institution nor with an apparatus; it is a type of power, a modality for its exercise, comprising a whole set of instruments, techniques, procedures, levels of application, targets; it is a 'physics' or an 'anatomy' of power, a technology" (215).

13. Foucault, *Discipline and Punishment*, 178.

14. Others argue that a feminist Foucauldian approach is misguided. See Nancy Hartsock, *The Feminist Standpoint Revisited & Other Essays* (Boulder: Westview Press, 1998); Somer Brodribb, *Nothing Mat(t)ers: A Feminist Critique of Postmodernism* (North Melbourne: Spinifex Press, 1992); and Rosi Braidotti, *Nomadic Subjects* (New York: Columbia University Press, 1994).

15. Sandra Lee Bartky, *Femininity and Domination: Studies in the Phenomenology of Oppression* (New York: Routledge, 1990). Bartky, to be clear, raises a concern regarding Foucault's theoretical model; namely, it lacks the necessary vocabulary to

formulate the nature and meaning of embodied resistance and agency (81). However, as I will show, her critique may be overstated.

16. Bartky, *Feminity and Domination*, 65.

17. Ibid.

18. Ibid.

19. Ibid., 66.

20. See Kim Chernin, *The Obsession: Reflections on the Tyranny of Slenderness* (New York: Harper and Row, 1981).

21. Foucault, *Discipline and Punishment*, 201.

22. Bartky, *Feminity and Domination*, 73.

23. Ibid., 74.

24. See Sondra Solovay, *Tipping the Scales of Justice: Fighting Weight-Based Discrimination* (New York: Prometheus Books, 2000).

25. Susan Bordo, *Unbearable Weight: Feminism, Western Culture and the Body* (Berkeley: University of California Press, 1993), 186.

26. Although Weight Watchers has spent millions in their attempt to attract men to the program, the majority (about 90 percent) of clients remain female. See E. J. Schultz, "Weight Watchers Pick a New Target: Men," *Advertising Age*, last modified April 22, 2011, http://adage.com/article/news/weight-watchers-picks-a-target-men/227155/.

27. Cressida Heyes, *Self-Transformations: Foucault, Ethics, and Normalized Bodies* (Oxford: Oxford University Press, 2007).

28. Foucault, *History of Sexuality*, 95, my emphasis.

29. Ibid., 93.

30. Foucault distinguishes between power and domination. Domination is static, asymmetrical and points to the ossified relations of powers. He cites conventional marriage in the eighteenth and nineteenth centuries as an example. See Michel Foucault, "The Ethics of the Concern for Self as a Practice of Freedom," in *The Essential Works of Foucault 1954-1984: Vol. 1: Ethics, Subjectivity and Truth*, ed. Paul Rabinow (New York: New Press, 1997), 293.

31. Margaret A. McLaren, *Feminism, Foucault, and Embodied Subjectivity* (New York: State University of New York Press, 2002), 41.

32. Foucault, *Discipline and Punishment*, 95.

33. Heyes, *Self-Transformations*, 87, my emphasis.

34. Ibid.

35. Judith Butler, "What is Critique? An Essay on Foucault's Virtue," in *The Political: Readings in Continental Philosophy*, ed. David Ingram (Oxford: Blackwell, 2002), 215.

36. Michel Foucault,"Qu'est-ce que la critique?" *Bulletin de la Societe Francaise de Philosophie* 84 (1990): 35–63, 39. Quoted in Heyes, *Self-Transformations* 117. Heyes's translation.

37. Heyes, *Self-Transformations*, 68.

38. Ibid.

39. Glenn A. Gaesser, *Big Fat Lies* (New York: Fawcett Columbine, 1996), 81.

40. Ibid., 5.

41. Michael Kimmel, "Masculinity as Homophobia: Fear, Shame, and Silence in the Construction of Gender Identity," in *Theorizing Masculinities*, ed. Harry Brod and Michael Kaufman (Thousand Oaks, CA: Sage Publications, Inc., 1994).

42. Ibid., 132.
43. Ibid.
44. Ibid.
45. Ibid., 138.
46. Jeffery Sobel, "Men, Meat, and Marriage: Models of Masculinity," *Food and Foodways* 13 (2005): 137. It is important to note that there are multiple models of masculinity with each interacting with other aspects of culture including one's class, race, ethnicities, sexual orientation, age and regions. While I maintain that the model presented here is the prototype or commonly idealized form of masculinity I also acknowledge the multifaceted and dynamic nature of multiple masculine scripts.
47. Carol Adams, *The Sexual Politics of Meat: A Feminist Vegetarian Critical Theory* (New York, Continuum, 1990), 34.
48. American Meat Institute, "The United States Meat Industry at a Glance," accessed June 15, 2014, http://www.meatami.com/ht/d/sp/i/47465/pid/47465.
49. Hank Rothgerber, "Real Men Don't Eat (Vegetable) Quiche: Masculinity and the Justification of Meat Consumption," *Psychology of Men and Masculinity* 14 (2013) and Matthew B. Ruby and S. Heine, "Meals, Morals, and Masculinity," *Appetite* 56 (2011): 447–50, respectively.
50. Marion Nestle, *Food Politics*: *How the Food Industry Influences Nutrition and Health* (Los Angeles: University of California Press, 2007).
51. Arthur Allen, "U.S. Touts Fruit and Vegetables While Subsidizing Animals that Become Meat," *Washington Post Health and Science*, October 3, 2011, http://www.washingtonpost.com/national/health-science/us-touts-fruit-and-vegetables-while-subsidizing-animals-that-become-meat/2011/08/22/gIQATFG5IL_story.html.
52. Sobel, "Men, Meat and Marriage," 139.
53. Nestle, *Food Politics*, 143.
54. The following examples are from Arran Stibble, "Health and the Social Construction of Masculinity in *Men's Health* Magazine," *Men and Masculinities* 7 (2004): 31–51.
55. *Men's Health*, September 2000, 90.
56. Ibid., 49.
57. Ibid., 166.
58. On Patrik Baboumian and Billy Simmonds, see Great Vegan Athletes, accessed March 23, 2014, http://www.greatveganathletes.com/vegan_athlete_patrik-baboumian-vegan-strongman and http://www.greatveganathletes.com/vegan_athlete_billy-simmonds-vegan-bodybuilder, respectively. On Arian Foster, see Kevin Gray, "Going Vegan in the NFL," *Men's Journal*, December 2012, http://www.mensjournal.com/magazine/going-vegan-in-the-nfl-20130123.
59. Kasey Panetta, "Should You Try a Vegan Diet?" *Men's Health*, June 8, 2012, http://new-mh.menshealth.com/nutrition/vegan-diet-training.
60. "Meat Consumption and Cancer Risk," The Physicians Committee on Responsible Medicine, accessed May 14, 2014, http://www.pcrm.org/health/cancer-resources/diet-cancer/facts/meat-consumption-and-cancer-risk.

61. Will Courtenay, "Behavioral Factors Associated with Disease, Injury, and Death Among Men: Evidence and Implications for Prevention," *The Journal of Men's Studies* 9 (2000): 89.

62. Ibid., 90.

63. American Cancer Society, "What are the Risk Factors for Prostate Cancer?" last modified February 25, 2014, http://www.cancer.org/cancer/prostatecancer/overviewguide/prostate-cancer-overview-what-causes.

64. Medical News Today, "Why Men Are More Prone To Heart Disease: New Research Led By University Of Leicester," September 1, 2008, http://www.medicalnewstoday.com/releases/119844.php.

65. Robert A. Koeth et al., "Intestinal Microbiota Metabolism of L-carnitine, a Nutrient in Red Meat, Promotes Atherosclerosis," *Nature Medicine* 19 (2013): 576–85, accessed July 13, 2014, doi: 10.1038/nm.3145.

66. D. A. Snowdon, R. L. Phillips, and G. E. Fraser, "Meat Consumption and Fatal Ischemic Heart Disease," *Preventive Medicine* 13 (1984): 490–500.

67. There are other masculine activities linked to early mortality (e.g., smoking, excessive alcohol consumption, occupations, fewer doctor visits). See R. F. Levant, R. Wu, and J. Fischer, "Masculinity Ideology: A Comparison Between U.S. and Chinese Young Men and Women," *Journal of Gender, Culture, and Health* 1 (1996): 207–20.

68. Physicians Committee on Responsible Medicine, "Meat Week Should Be Renamed Erectile Dysfunction Acceptance Week, say Doctors," accessed July 14, 2014, http://www.pcrm.org/media/news/meat-week-be-renamed-erectile-dysfunction-week.

69. P. Tharyan and G. Gopalakrishanan, "Erectile Dysfunction," *Clinical Evidence* (2006): 1803, accessed June 12, 2014, http://www.ncbi.nlm.nih.gov/pmc/articles/PMC2907627/.

70. Heyes, *Self-Transformations*, 70.

71. Chole Taylor, "Foucault and the Ethics of Eating," *Foucault Studies* 9 (2010): 80.

72. Marilyn Frye, *The Politics of Reality: Essays in Feminist Thought* (California: The Crossing Press, 1983), 10.

Chapter 10

Food, Film and Gender

Margaret Crouch

Each human being inhabits what I call a "food world." A food world encompasses the meanings that food and the activities surrounding food have for a person. These meanings are cognitive, emotional and sensory. They are constructed by, and construct, our particular relationships to eating and to food. These relationships are shaped by our experiences, which in turn are influenced by biology and culture. The influence of biology is evident in studies that have found that at least some of our taste experiences are determined by our genes. People often report that certain vegetables, such as broccoli, taste bitter, and that eating them is unpleasant. Tasting broccoli as bitter has been associated with the possession of a particular gene.[1] The influence of culture on our relationship to food has been studied by anthropology under the concept of *foodways*: "The beliefs and behaviour surrounding the production, distribution, and consumption of food."[2]

Food worlds include the amount of mental and emotional space food and eating consume. Some of us think about food a great deal, others hardly at all. Some of us think about food, but only in a general way, without much attention to the kind of food. Others think about food *only* in terms of kind of food. Most are somewhere in between. Dieters sometimes seem to think about little but food. Because food and eating involve sensation, cognition and emotion, food worlds also include our thinking and feelings about our bodies, and the relationship between our bodies and food. Our food worlds include how we react to feelings of hunger, as well as who we think is responsible for feeding us, and who is responsible for cleaning up after they have fed us.

Though each of us has our own food world, there are discernable patterns that link these worlds according to group membership. Some groups are formed biologically, perhaps according to whether or not they possess a particular gene. Others are formed culturally, according to gender, or ethnicity, for example. I am interested here in the gendered patterns evident in the food worlds depicted in film. In this chapter, I will discuss some of the ways in which food and gender are portrayed in film, arguing that, in the world of film, women and men inhabit different food worlds: their pleasures in food, their anxieties about food, their intimate relationships with food, and their control of food, differ, revealing the way that in mainstream film, food, as so much else, is ruled by heteronormativity. Food and eating in film are so gender stereotyped that violating the norms of those gendered stereotypes always communicates something about the gender of the character doing the violating, as I shall demonstrate below. Because mainstream Hollywood romances and comedies most clearly portray what is considered natural and normal for heteronormativity generally, most of my examples will be taken from such films.

I. GENDER AND FOOD

Much has been written about women and food; somewhat less has been written about men and food, though that is slowly changing. Susan Bordo's *Unbearable Weight*, first published in 1993, was one of the first books to set out a philosophical perspective on the relationship between gender and food. She argues that the history of Western philosophy, which privileges the mind over the body, identifies men with mind and women with body. The mind and its rationality are what distinguish us from nonhuman animals. The animal body is necessary for life, but its appetites should be governed by the mind. Women, considered less rational and more driven by appetites and desires, are thought to have more trouble controlling their appetites than men. Perhaps as a consequence, heteronormativity requires that women be hypervigilant about what they eat, that they refrain from expressing hunger or the desire for food, and that they maintain bodies that are slim and youthful, and attractive to heterosexual males. In medical circles, the kind of eating typical of women is called "restrained eating": "Restrained eating refers to a persistent pattern of eating-related cognitions and behaviours in order to reduce or to maintain body weight."[3] As we shall see, women in mainstream film, especially if they are romantic leads, are commonly depicted as restrained eaters.

Sandra Bartky describes the effort involved in controlling hunger and appetite as a "discipline" of heterosexual femininity.[4] From a very young age, many women learn to monitor their food intake and the appearance of their bodies. Recent studies show that one out of every four girls of fifteen is dieting.[5] Other studies show that "from approximately 6 years of age, many girls both desire a thinner ideal body and are aware of dieting as a means to achieve this."[6] Furthermore, exposure to media influences, such as television and adult magazines, predicts which young girls wish to be thinner, and consider dieting as a means to that end.[7] The constant monitoring of one's food intake is part of the food worlds of many women.

Though food worlds vary with culture, and the heteronormativity I have been describing is based on the West, the concern with what one eats and how much is not limited to the West. Concerns with eating and body image very similar to those in the West are common in Japan, China and South Korea, as well as other countries. [8] The thin ideal for girls and women is most influential among higher socioeconomic populations, and seems to increase as a culture becomes wealthier. South Korea has become one of the cosmetic surgery capitals of the world. It is not only South Korean women who undertake surgery in South Korea, but also women from other Asian countries. The most common cosmetic procedure, worldwide, is fat removal, followed by breast augmentation.[9] A much-cited set of studies of women in Fiji show that prior to the introduction of television, the ideal body type for women in Fiji was substantial, and thinness was ridiculed. After three years of exposure to *Baywatch*, thinness replaced the previous ideal body type, and young women began exhibiting eating disorders, which had never been a problem among Fijians before.[10]

As Bordo reported in the 2003 edition of her book, some of the anxiety over body image and related relationships to food typical of women are spreading to men. For men, the ideal body type has become the lean, muscular body displayed in action films.[11] This body requires fuel and a great deal of exercise to create and maintain. The idea of food as fuel is common in relation to men. Truck stops, for example, provide fuel for both the truck and the driver. In a Norwegian study, a truck driver expressed a food world common to many men:

> *D1:* Food actually means? But I have no relation, I'm no gourmet, I'm more of a gourmand, if you know the difference?
>
> *I:* It's the amount?
>
> *D1:* Because I eat to give nourishment to the body. Just like filling up the tank with diesel.
>
> *I:* Like to have energy to drive?
>
> *D1:* Yes. You are hungry. And I have never been a, I have eaten both Russian caviar and goose liver paté, both are tragic things. But food is simply something you have to be able to survive.[12]

More gay men than heterosexual men are concerned with body image. Such men tend to be more concerned about muscularity than weight.[13] One study found that, in a comparison of lesbian women, heterosexual women, gay men and heterosexual men, heterosexual men "reported more positive evaluations of their appearance, less preoccupation with their weight, more positive effects of their body image on their quality of life and the quality of their of their sex life."[14] The study found little difference between lesbian women and heterosexual women with regard to body dissatisfaction. A meta-analysis of studies of body image published in 2004 found that gay men seemed slightly less satisfied with their bodies than heterosexual men, and lesbians very slightly more satisfied with their bodies than heterosexual women.[15]

Traditionally, men have been allowed to express hunger and appetite. Many men are more interested in the amount of food available to them, rather than kind of food. Will they get enough? When will they be able to eat again? To pay too much attention to what one eats was once considered unmanly. These days, a man may prepare food and shop for food without diminishing his masculinity—perhaps due to the "professionalization" provided by men in food television.

Men and women still have different relationships to the planning and preparation of food in their households. Most domestic food preparation in the United States is still done by women. When women are wealthy enough, they often hire women of lower status to perform these tasks for them. According to the American Time Use Survey of 2013, on average, 68.1 percent of women spent an average of 1.17 hours of their time each day on food preparation and clean up, and 16.9 percent spent an average of .81 hours per day shopping for groceries; on average, 41.7 percent of men spent an average of .80 percent of their time each day on food preparation and clean up, and 14.2 percent of men spent an average of .65 percent of their time grocery shopping. There is evidence that men are doing more of the work shopping for and preparing food in the home than in the past.[16] It is also the case that more people are dining out than in the past: "The percent of food expenditures for food eaten outside the home increased from 33 percent in 1970 to 49 percent in 2005."[17]

When males prepare food, it is primarily for special occasions at home, or commercially as chefs or bakers. Women constitute 40.6 percent of cooks, but only 21.4 percent of "chefs and head cooks."[18] These statistics do not distinguish kinds of cooking and food preparation. Men are often responsible for barbequing or deep frying turkeys—forms of cooking that take place outside, and involve open flame.

But it is the emotional world of eating that seems most to differentiate the food worlds of men and women under dominant conceptions of femininity and masculinity. Many women express love through feeding children, friends, or romantic partners. My own mother's first words, when one of her children or grandchildren comes through her door are, "Would you like something to eat?" For many women, food and eating are fraught with anxiety, even as eating particular foods is a method for allaying anxiety or negative emotions. Starving oneself can also provide feelings of the sublime, as Sheila Lintott argues in "Sublime Hunger: A Consideration of Eating Disorders beyond Beauty." Lintott beautifully describes the food world of a woman with an eating disorder:

> You wake up at 5 A.M., dizzy, with an empty feeling in the pit of your gut. Your first thoughts are of food, but not in any simple sense. Instead of thinking about some delicious meal that might satisfy your hunger, you think quite the opposite. You think that today you will not eat until 5 P.M. or 6 P.M., or, best of all possibilities, not at all. You deliberate, figuring when you will have to eat, and how you will be able to avoid eating until then, without detection. Today, you affirm, as you do every day, that you will eat less than yesterday. Before falling asleep last night, while doing your sit-ups in bed, you already made a plan to run five extra miles this morning to make up for the potato you ate yesterday. You are guiltily aware that you were not supposed to eat that potato; you know you should have eaten only some celery. You know that if you eat, you may lose control and devour more food than most people eat in a week. But you find comfort in your confidence that if this happens, you can deal with it; you can vomit it up. You know the tricks—how to make yourself vomit, silently and quickly if need be.[19]

For men, the most common emotions related to food seem to be anxiety about having access to food and getting enough food, and positive emotions. This quote from Hemingway's *A Moveable Feast* expresses a typical masculine attitude toward food:

> As I ate the oysters with their strong taste of the sea and their faint metallic taste that the cold white wine washed away, leaving only the sea taste and the succulent texture, and as I drank their cold liquid from each shell and washed it down with the crisp taste of the wine, I lost the empty feeling and began to be happy and to make plans.[20]

The role of emotion in food worlds has been examined empirically from a variety of perspectives. Much of this research focuses on the effects of emotion on eating behavior, as a way to understand eating disorders and obesity. For example, Michael Macht identifies "five classes of emotion-induced changes of eating: (1) emotional control of food choice, (2) emotional suppression of food intake, (3) impairment of cognitive eating controls, (4) eating to regulate emotions, and (5) emotion-congruent modulation of eating."[21] As this classification suggests, our relation to food is deeply affected by emotion; but the relationship between particular emotional states and amount of food intake, or the kind of food eaten, varies dramatically for different individuals. In general, "high-arousal or intense emotions suppress eating, and negative emotions can increase or decrease food intake."[22] Macht argues that the eating of restrained eaters, emotional eaters and normal eaters is affected differently by the same emotion. These different kinds of eaters are determined by their placement on a scale. Emotional eaters are people who use "food to modify negative mood states."[23] For example, "negative emotions increase food intake in restrained eaters as well as intake of sweet-high-fat food in emotional eaters."[24] Macht suggests that restrained eaters eat more in response to negative emotions because these emotions interfere with their cognitive control of eating—they distract them and they "lose control." Normal eaters sometimes engage in emotional eating, but unlike emotional eaters, who routinely eat in response to negative emotions, this behavior is transient.

Macht does not say which genders tend to be which kinds of eaters. However, if we look at the association of emotion with women in Western tradition, we can predict that non-normal eaters will tend to be female, and normal eaters male. Restrained eaters try to control their food intake, and this control can be disrupted by strong emotions. Emotional eaters use food to allay strong negative emotions, that is, to control them. Emotions generally play a larger role in the lives of non-normal eaters, so it is not surprising that they play a large role in their relation to food and eating. We need only think of the stereotype of the disheartened single woman eating ice cream from its container in her kitchen late at night—a scene in numerous television programs and movies. This is emotional eating *par excellence*.

Men do engage in emotional eating, but there is evidence that they eat in response to positive rather than negative emotions: "Men may be more likely to eat comfort foods to maintain or enhance *positive* emotions, whereas emotional eating in women tends to be driven by *negative* mood states."[25] Interestingly, the kinds of food that men eat to maintain positive emotions are not the same as those that women eat to alleviate negative emotions: "Men experienced the most intense positive affects prior to consuming foods with low calorie content, whereas women experienced more negative affects particularly before consuming high-calorie sweet comfort foods."[26]

Men have much less emotional attachment to food and eating than women, though they do use food to express their identities as men.[27] Foods are gendered, and men use this to perform hegemonic masculinity. "Meats are masculine, vegetables feminine. The steak is probably the most masculine food in our society. . . . The nonmeat salad is the epitome of femininity."[28] White wine is feminine, red less so, but beer is more masculine than wine. "Men are considered to be less likely than women to avoid fat, eat fibre, eat fruit and diet, and attached less thought to healthy eating. . . . Red meat . . . has been found to represent for men a totem of virility and strength."[29] Men are not limited to hegemonic forms of masculinity, and use food choice to express their own versions of masculinity. In addition to food choice, the amount of food eaten is gendered. "Traditionally, men are assumed to possess a voracious appetite."[30] However, one group of men surveyed (in Ireland) saw the sensual enjoyment of food as feminine—it suggests the submission of oneself to emotion and appetite. "When they do derive enjoyment from food, it is only occasionally, and in social situations most typically when they are seated as head of the family dinner table. . . . There is also a preference for meat products, large portions, and the use of alcohol during meals."[31]

II. FOOD, SEX AND GENDER

Appetite for food has been paired with appetite for sex since the beginnings of Western philosophy. As Carolyn Korsmeyer points out, "Appetite is often conceived as a twin drive for food and sex."[32] As we saw earlier, women are considered more vulnerable to appetites. This is because the Western philosophical tradition is written from the perspective of heterosexual males, who maintain a hierarchy of senses according to the perceived proximity of the sense to the body. Sight is associated with knowledge; its objects are considered to be objective and to preserve a distance from the body. Because of this, sight is associated with masculinity, and has been considered a worthy subject of philosophical contemplation. Taste, on the other hand, is closely associated with the body. Taste does not occur without contact between a body and its object. Taste is also considered subjective, and, thus, is associated with femininity. According to Korsmeyer,

> Implicit in Plato's view is the belief that the ability to transcend the body, to govern the senses, to gain knowledge, is a masculine ability that when exercised well will keep one embodied as a male. There is, therefore, an implicit gendering of the use of the senses themselves, with the higher, distal senses of sight and hearing paired up with the controlling intellect of a virtuous man, and the lower, proximal senses with the appetites and the dangerous pleasures that are in one way or another associated with femininity.[33]

The hierarchy of the senses continues in European painting. The association of masculinity with the "higher" senses and femininity with the "lower" is also retained. Women's bodies are associated with food and sex: "A common denominator in all associations of female bodies and edibles is the ambiguous meaning of 'appetite,' which connotes both sexual and gustatory craving for satisfaction, an association that appears to be more or less universal across dramatically different societies."[34] As the ideal knower is heterosexual and male, so is the ideal viewer of painting.

The association of eating and sex, and of both with women, found in Western philosophy and the arts, is also found in film, as I shall show below.

III. FOOD AND FILM

Food may often appear in films, but it is rarely eaten by any character. In *Food in the Movies*, Steve Zimmerman explains that for many years, depicting food was impractical, since the intense, hot lighting required to shoot a scene would destroy the food. Even now, the need for repeated takes makes depicting a meal onerous, as the food must be prepared and placed in the same way over and over again. There are also aesthetic reasons for not depicting eating. Eating rarely moves the plot along, and when done properly, is uninteresting to watch. Thus, the food in films plays various roles, but is almost never simply the everyday act of eating.

The fact that food is not often eaten in film means that food is usually "background." We hardly notice it—it appears "natural," and does not call attention to itself. This enables us to examine this very "naturalness" to determine what is considered "normal" with regard to food, eating and gender. The less remarkable the food or eating in a scene, the more likely that it is making use of norms of gender and food in society. This point is made by Fabio Parasecoli:

> Food-specific scenes, due to their secondary role, are perceived as natural and normal, thus becoming virtually invisible to the viewers. This invisibility allows actors, scriptwriters, and filmmakers to display cultural elements deriving from widespread models of masculinity that in other, more relevant or spectacular parts of the movies would not be expressed so freely. . . . Whatever the reason for the presence of food scenes, their apparent ordinariness and familiarity offers an apt environment for the representations of values, attitudes and behaviors that reflect widely accepted and culturally sanctioned templates of what a man should be like and act like.[35]

Think about the appearance of beer cans—empty or full—in a house. Or, seeing an open refrigerator with small containers of flavored yogurt and a bottle of white wine. We assume immediately the gender of the person who inhabits that domicile.

Often food acts as a prop, an excuse for people to get together in one place and for relationships to be revealed. Movies about families commonly have many meal scenes, where we come to see how the members of the family relate to one another. For example, Woody Allen's *Hannah and Her Sisters* has numerous family dinner scenes, each one more fraught than the next.[36] Revealing secrets, exposing tensions and depicting conflicts at a dinner table is so common that when one sees the characters preparing to eat together, one's anxiety mounts, in expectation of some sort of conflict. These tensions and conflicts can begin even before the dining scene, with invitations to a meal. Another family movie, *The Kids Are Alright*,[37] also contains many scenes around the dinner table. In the first, we are shown a great deal about the two mothers and their relationship. Later, the biological father of the children is invited to dinner. We see the relationships between the man and the children, as well as the conflict he is introducing into the relationships among all of the characters, in the course of the dinner.

Food also serves as a "time-compression device" in film.[38] The preparation of food can be seen—several women in a kitchen, for example—then the cleanup. Food can serve to reveal features of a character. For example, in *The Upside of Anger*, a boyfriend of the daughter comes to dinner at the mother's house.[39] He slurps his soup, revealing his class background, thereby showing his inappropriateness as a serious romantic interest for the daughter.

IV. GENDER AND FOOD IN FILM

In mainstream Hollywood films, as well as in many independent films, women and men inhabit different food worlds. The gendered stereotypes about food and eating that appear in such films aid in establishing for the audience that the character does not challenge heteronormative norms. Women in such films rarely eat; they often prepare food in domestic contexts, less frequently professionally. When women are shown eating, it is usually tiny amounts of low-fat or low-carb food. If they eat something else—in most cases, something sweet—they often comment on the eating, justifying it, or expressing guilt about eating it. Such eating is depicted as problematic and exceptional. They are not shown as taking pleasure in eating. While eating a chocolate might be enjoyed in the moment, the guilt and anxiety that follow undermine that momentary pleasure.

As I mentioned earlier, this kind of eating is called "restrained eating."[40] Thus, in Hollywood films, leading ladies are most often portrayed as restrained eaters. The food world of the restrained eater is complex, and food dominates much of her thinking. It is difficult for such a person to eat without anxiety. In *The Holiday*, Amanda (Cameron Diaz) describes the kind of holiday she wants. Included in her litany of desires is "I want to eat carbs without wanting to kill myself."[41] This is illustrative of the food world of the restrained eater. She is always calculating and balancing her food intake. Is the dessert worth the calories? If she has a salad, then she can have a dessert.

She will eat an appetizer, but not an entrée. Has she had anything highly caloric yet today? Is today a gym day, where she can be sure to burn off some of the calories she takes in? These calculations are ongoing.

Even though it has long been thought that African American women have a broader range of acceptable body types, this seems to be changing. Cheryl Thompson demonstrates that, though there is a different narrative surrounding the weight of African American women, the tyranny of weight control affects them as much as any other ethnic or racial group. This is indicated by the number of African American women who have advertised weight loss programs and products in recent years. Queen Latifah advertised for Jenny Craig, Jennifer Hudson for Weight Watchers and Janet Jackson for Nutrisystems.[42] This shift is evident in Hollywood films featuring African American women in romantic leads. Though some actresses are curvier than their white counterparts, they are usually very slim.

As Thompson points out, the weight of African American women in films, as in life, is often associated with their class. In many romantic films starring African American female leads, the slim and fit women are from upper-class families and have successful professional lives. For example, in *Something New*, the lead actress, Sanaa Lathan, is slim and trim, as are all of her friends.[43] In Tyler Perry's *The Single Mom's Club*, an African American mother from a lower socioeconomic class than most of the other women in the group carries more weight.[44] However, unlike in many films, this woman does have a class-appropriate romantic partner.

The restrained eating of female characters, especially at restaurants, is often occasion for comedy. In *Shallow Hal*, the male lead, played by Jack Black, is hexed to only see the inner beauty of people.[45] He meets a woman, Rosemary, who appears slim to him, but overweight to the audience. Rosemary, in both incarnations, is played by Gwyneth Paltrow. Hal and Rosemary are eating in a restaurant, and Rosemary orders a cheeseburger, chili fries and a milkshake. What comes next is a variation on one of the most common lines in movies about women ordering in restaurants:

> It's nice to see a girl order a real meal; I can't stand it when you guys order a glass of water and a crouton—it ruins the whole point of going out!

The contradictoriness of this sentiment is stunning. Men in the Hollywood film world, even those like Hal, will not consider dating women who are not slim and beautiful. But most women cannot eat like this and remain slim and beautiful. Rosemary is evidence of this.

If a leading woman character is not a restrained eater, she is defeminized or pathologized. The defeminization may take the form of a combination of unrestrained appetites for food and sex. This is allowed, but typically only under very specific circumstances. One such is at certain points in heterosexual relationships, especially the most sexually active parts. The pathologization of nonrestrained eaters occurs an independent film entitle *Disfigured*.[46] The film portrays an overweight woman who is marked as psychologically troubled. In one scene, she sits down to a brimming plate of food in her apartment, binging on fatty foods. Even more disturbing in their depiction of overweight girls as psychologically deranged are the Australian film *In Her Skin*[47] and the French film *Fat Girl*.[48] In the former, a young girl is killed, and the killer is an overweight girl who wishes to be the dead girl. In the latter, an overweight young girl exhibits psychologically problematic behaviors in relation to her older sister's sexual relationship. Both girls are also shown as rebellious and disobedient to their parents; however, this behavior is not shown as psychologically healthy at all.

There are a very few actresses who are powerful enough not to be thin, restrained eaters in leading Hollywood roles. One of these is Queen Latifah. However, even she has been portrayed as a restrained eater. She appears as the lead in a film illustrating a familiar exception to women restrained eaters: the woman who faces death.

In *Last Holiday*, Queen Latifah plays Georgia, a woman restrained and pleasure denying in every aspect of her life.[49] She makes gourmet meals following the instructions on television food shows, but she never eats them. She gives them away and eats Lean Cuisine. Upon receiving a diagnosis that gives her just weeks to live, she takes all of her money and goes to a hotel in the European Alps where one of her favorite chefs heads the kitchen. She begins to enjoy herself, taking pleasure in spending her money on a luxurious suite and ordering every special on the menu at dinner. Her attitude toward eating and pleasure is not resolved at the end of the film when she learns that the diagnosis was in error. However, she does have a boyfriend and has opened herself to pleasure and appetite.

Another example of this exception to the restrained eater appears in the film *Seeking a Friend for the End of the World*.[50] Earthlings have learned that everyone will die in two weeks when the earth is hit by an asteroid. At a dinner party, guests are asked how they will spend the remaining days of their lives. One of the guests, a single woman—portrayed as silly and vapid—gushes that one of the things she will do is "eat whatever I want, and not even care." The audience is thus informed that her food world is that of the restrained eater.

In addition to restrained eaters, women are often portrayed as emotional eaters much more frequently than men. As we saw above, women are more likely to eat in reaction to negative emotions. The scene of the single woman in her kitchen, at night, with a container of ice cream and a spoon, is a cliché in romantic comedies. Another aspect of emotional eating is secret eating. A woman can pretend to everyone around her that she is eating healthily, but the call of sweets may be too strong to resist. This kind of emotional eating is so identified with women that when men do it, they are feminized. In *This is 40*, Pete (husband) eats cupcakes in secret, while his wife Debbie smokes in secret.[51] Cupcakes are already identified as feminine food, so his eating them in secret displays the feminization of his character. His feminization is reinforced by other scenes in the film. When Pete rides his bicycle into an opened car door and ends up in the hospital, Debbie and father are in the hall waiting for him. They have a discussion in which the father tells Debbie that she is the fighter in the relationship, further emasculating the male character.

The relationship of eating, women and sex is complex in film. The meanings of these concepts, alone or in combination, is so complex that contradictions abound. The association of sex, food and the feminine is more or less explicit in film, especially in mainstream Hollywood films; as is the case for so many social norms, those relating food to gender are often both expressed and ironized simultaneously. For women in films, food is often a substitute for sex, or for intimacy. A woman has *either* food or sex. This carries through to the rarity of romantic relationships for overweight women in films. Overweight women apparently do not have sex—they have food. Of course, there are some exceptions. In a very few cases, the star power of an actress is sufficient to allow for a romantic relationship. I have already mentioned *Last Holiday*. Another exception is Melissa McCarthy. In *Tammy*, she is married, and then finds a much better romantic relationship in the course of the film.[52]

Sometimes thin, attractive women do eat in films. When they do, they are almost always also having a lot of sex. This is true in the two *Sex and the City* films.[53] The four main characters are often eating, and often having sex. But sometimes, eating substitutes for sex. In one scene, Samantha has been eating (and has a belly to show for it) to avoid having sex with her neighbor, thereby cheating on her boyfriend. However, overweight women who eat a lot typically are not seen as sexually attractive. In a sense, such women are defeminized—they are not the objects of sexual desire by men. Lydia, the overweight woman in *Disfigured,* does have sex with a man, but it is unsatisfying and she ultimately confronts him about using her. We might say that the norms of femininity require "restrained sex" *and* "restrained eating" of women. Women are not supposed to be lustful, except under very specific conditions, one of which is when she is in the most passionate part of a heterosexual romantic relationship. Perhaps the food is burned off by the sex. Interestingly, often when women exhibit appetites for sex and food, they are masculinized. This is very evident in portrayals of female vampires. They are unrestrained with regard to eating, sex and violence. They are simply unrestrained, and they are threatening.

For men in films, food is not a substitute for sex. Rather, it is in addition to sex. Films featuring overweight men do not portray them as having *either* food or sex. Rather, they get both. Men do not deny themselves either one. Overweight men, in films, can be sexually active, and attractive to women. In *A Big Love Story*, a man weighing over 400 pounds begins a regime of dieting and working out to try to lose weight.[54] He has a personal trainer. Eventually they begin dating. The man has lost a bit of weight, and is in better shape, but a true transformation of his body is in the distant future.

Men in films are not shown as eating much, either, but not because eating would violate the norms of masculinity. For the lean, muscled men of action films, food is just a necessary part of life, like sleeping—which is not shown, either. Food is fuel, and the point is to eat a lot of it and not spend much time thinking about it. In fact, we never see them eat, which makes us wonder how they keep up the fighting and risk taking activities required of action heroes.

When men do pay attention to the food they eat, their masculinity is called into question. This poses problems for male characters in food films, films that are designed around food. However, to save their masculinity from the threat, they are professionals rather than merely domestic cooks. As in other spheres of life, in film, the food worlds of male and female cooks are very different.

When men are cooks in film, there is often competition involved in their cooking. Cooking is about ego. In the film *Chef*, a restaurant chef is devastated by a critical review.[55] As a result, he quits the high-stakes professional chef life and begins making cubanos, meaty sandwiches that are delicious and can be sold out of a food truck. The reviewer provides a positive review of the cubanos, and our chef's damaged ego is made whole.

The competitive nature of cooking is often depicted as a competition between fathers and sons. For example, many young men portrayed as chefs in film have fathers who are also chefs. They try to meet their fathers' expectations of manhood by competing with their fathers and winning their respect. This is often done with an ethnic flavor. In a number of films, the father has an ethnic restaurant or bakery. The son must make his own way, but eventually returns to his father's ethnic roots to make his food authentic. This resolves the conflict between the son and his father. This is true of the film *Today's Special*.[56]

The tendency to emphasize ethnicity in films featuring male cooks can be seen in Ang Lee's *Eat, Drink, Man, Woman* and the U.S. version, *Tortilla Soup*.[57] Both feature families with fathers but no mothers, and daughters but no sons. The fathers own high-end restaurants and cook for their daughters weekly at large, complex family meals. As I said earlier, a meal is usually not about eating. This is certainly true in these films. The fathers are losing their sense of taste, and the daughters do not want to be at these dinners.

The fathers are losing the daughters to their own lives. In both narratives, one of the daughters is a cook, and she eventually begins cooking for her father, which restores his sense of taste—like magic. The fathers want to express love for their daughters through their cooking, but are unable to do so, until one of the daughters enables it through her own offering of love.

When women are professional chefs or bakers, they often begin the narrative as masculinized by this professionalization of a woman's domestic role. In *Mostly Martha* and its U.S. remake, *No Reservations*, a female chef is controlling and aggressive with regard to the food that comes out of her kitchen.[58] Her boss sends her to an analyst because of her hostility toward patrons who criticize her food. The analyst is not able to do much. It is only by becoming more feminized that she is able to lose these negative (and masculine) characteristics. This happens when she must take over the rearing of her sister's child, and when a male chef and love interest is introduced into her kitchen. The heteronormativity of this narrative is striking. A woman can only be a professional chef if she is also in a heterosexual relationship and has a child. This then makes her professional cooking an extension of domestic care for her loved ones, which is as it should be.

That a woman's professional food preparation is an extension of a mother's love is emphasized in the large number of films in which female chefs, bakers, or candy makers infuse their cooking with good magic, the result of which is to influence the emotions of those who eat the food. This is true of *The Mistress of Spices*, *Chocolat*, *Woman on Top*, *The Recipe*, *Like Water for Chocolate* and *Simply Irresistible*.[59] A twist on this is when a woman puts all of her emotion and passion into her cooking, as in the film *Waitress*.[60] There is nothing supernatural here, but the voice-over reveals the inspirations for her heavenly pies. "Bad Baby Pie" is created in response to finding out about an unwanted pregnancy. The inspiration for "I Hate My Husband Pie" is obvious.

V. CONCLUSION

I have demonstrated that the food worlds for women and men differ in reality and in Hollywood mainstream film. The lead female character in a Hollywood film is nearly always a restrained eater—unless she is a vampire. The lead heterosexual male character in a Hollywood film must not pay too much attention to food—unless he is a spectacular chef. Male characters prepare food competitively, to feed their egos. Female characters prepare food to provide care for others, to express their emotions, or to allow others to feed off their emotions.

Both food itself and a character's relationship to food are gendered. In order to maintain heteronormativity, a male character displaying a feminized relationship to food must be remasculinized, and a female character displaying a masculinized relationship to food must be refeminized. This is often done in film by, for example, providing a female chef with a child and a male romantic partner. For men, the masculinization can come from coming out on top in a cooking competition, or by reconnecting with his father, thereby taking his place as the head of the family business.

This is, of course, a very general survey of gender, film and food; and it is limited primarily to mainstream films and to films made in the United States. However, there are many popular films in Korea and China that mirror these same gendered relationships to food. And, wherever films are made, there is a pattern of masculine and feminine foods and food worlds upholding heteronormativity in popular entertainment.

NOTES

1. The gene is the *Ta2r38* gene. Rebecca Jacobson, "The Bitter Taste of Genetics," *PBS NewsHour*, December 23, 2010, accessed April 22, 2015, http://www.pbs.org/newshour/updates/science-july-dec10-geneticstaste_12-23/.

2. Carole M. Counihan, *The Anthropology of Food and Body: Gender, Meaning, and Power* (New York: Routledge, 1999), 2.

3. Michael Macht, "How Emotions Affect Eating: A Five-Way Model," *Appetite* 50 (2008): 2.

4. Sandra Lee Bartky, *Femininity and Domination: Studies in the Phenomenology of Oppression* (New York: Routledge, 1990), 63–82.

5. H. Sweeting and P. West, "Gender Differences in Weight Related Concerns in Early to Late Adolescence," *Journal of Epidemiology and Community Health* 56 (2002): 700–701.

6. Hayley K. Dohnt and Marka Tiggemann, "Body Image Concerns in Young Girls: The Role of Peers and Media Prior to Adolescence," *Journal of Youth and Adolescence* 35 (2006): 142.

7. Ibid., 141–51.

8. Michelle Lelwica, Emma Hoglund and Jenna McNallie, "Spreading the Religion of Thinness from California to Calcutta: A Critical Feminist Postcolonial Analysis," *Journal of Feminist Studies in Religion* 25 (2009): 19–41; Susan Bordo, *Unbearable Weight: Feminism, Western Culture, and the Body*, Tenth Anniversary Edition (Berkeley and Los Angeles: University of California Press, 2004); Kathleen M. Pike and Amy Borovoy, "The Rise of Eating Disorders in Japan: Issues of Culture and Limitations of the Model of 'Westernization,'" *Culture, Medicine, and Psychiatry* 28 (2004): 493–531; Naomi Chisuwa and Jennifer A. O'Dea, "Body Image and Eating Disorders amongst Japanese Adolescents: A Review of the Literature," *Appetite* 54 (2010): 5–15; Jie Yang, "*Nennu* and *Shunu*: Gender, Body Politics and the Beauty Economy of China," *Signs* 36 (2011): 333–57; Xiaoyan Xu, David Mellor, Melanie Kiehne, Lina A. Ricciardelli, Marita P. McCabe, and Yangang Xu, "Body

Dissatisfaction, Engagement in Body Change Behaviors and Sociocultural Influences on Body Image among Chinese Adolescents," *Body Image* 7 (2010): 156–64; Viren Swami et al., "The Attractive Female Body Weight and Female Body Dissatisfaction in 26 Countries across 10 World Regions: Results of the International Body Project I," *Personality and Social Psychology Bulletin* 36 (2010): 309–25; M. Makino, M. Hashizume, K. Tsuboi, M. Yasushi, and L. Dennerstein, "Comparative Study of Attitudes to Eating between Male and Female Students in the People's Republic of China," *Eating Weight Disorder* 11 (2006): 111–17.

9. Kathy Chu, "Extreme Dieting Spreads to Asia," *USA Today*, March 29, 2010; "Plastic Makes Perfect," *Economist Online*, January 30, 2013, accessed April 22, 2015, http://www.economist.com/blogs/graphicdetail/2013/01/daily-chart-22; Jung Ha-Won, "South Korea's Plastic Surgery Fad Goes Extreme," *Jakarta Globe*, May 27, 2013, accessed April 22, 2015, http://www.thejakartaglobe.com/features/south-koreas-plastic-surgery-fad-goes-extreme/?doing_wp_cron=1373226815.1865510940551757812500.

10. Anne E. Becker, "Television, Disordered Eating, and Young Women in Fiji: Negotiating Body Image and Identity during Rapid Social Change," *Culture, Medicine and Psychiatry* 26 (2004): 533–59; Anne E. Becker, Rebecca A. Burwell, Kesaia Navara, and Steven E. Gilman, "Binge Eating and Binge Eating Disorder in a Small-Scale, Indigenous Society: The View from Fiji," *International Journal of Eating Disorders* 34 (2003): 423–31; Anne E. Becker, Rebecca A. Burwell, Stephen Gilman, David B. Herzog, and Paul Hamburg, "Eating Behaviours and Attitudes following Prolonged Exposure to Television among Ethnic Fijian Adolescent Girls," *British Journal of Psychiatry* 180 (2002): 180.

11. Todd G. Morrison and Marie Halton, "Buff, Tough, and Rough: Representations of Muscularity in Action Motion Pictures," *The Journal of Men's Studies* 17 (2009): 68. This is true for both heterosexual and gay men.

12. Gun Roos and Margareta Wandel, "'Because I'm Hungry, Because It's Good, and to Become Full': Everyday Eating Voiced by Carpenters, Drivers, and Engineers in Contemporary Oslo," *Food and Foodways* 13 (2005): 172–73.

13. Maurice J. Levesque and David R. Vichesky, "Raising the Bar on the Body Beautiful: An Analysis of the Body Image Concerns of Homosexual Men," *Body Image* 3 (2006): 45–55.

14. Letitia Anne Peplau, David A. Frederick, Curtis Yee, Natalya Maisel, Janet Lever and Negin Ghavami, "Body Image Satisfaction in Heterosexual, Gay and Lesbian Adults," *Archives of Sexual Behavior* 38 (2009): 713.

15. Melanie A. Morrison, Todd G. Morrison, and Cheryl-Lee Sager, "Does Body Satisfaction Differ between Gay Men and Lesbian Women and Heterosexual Men and Women? A Meta-Analytic Review," *Body Image* 1 (2004): 127–38.

16. American Time Use Survey, 2013. Table A-1, http://www.bls.gov/tus/.

17. Patricia Allen and Carolyn Sachs, "Women and Food Chains: The Gendered Politics of Food," *International Journal of Sociology of Food and Agriculture* 15 (2007): 7.

18. United States Bureau of Labor Statistics, 2014, http://www.bls.gov/cps/cpsaat11.htm.

19. Sheila Lintott, "Sublime Hunger: A Consideration of Eating Disorders Beyond Beauty," *Hypatia* 18 (2003): 65.

20. Ernest Hemingway, *A Moveable Feast* (New York: Scribner, 2009), 18.

21. Macht, "How Emotions Affect Eating," 1.

22. Ibid., 4.

23. Robert D. Levitan and Caroline Davis, "Emotions and Eating Behaviour: Implications for the Current Obesity Epidemic," *University of Toronto Quarterly* 79 (2010): 783.

24. Macht, "How Emotions Affect Eating," 7.

25. Levitan and Davis, "Emotions and Eating Behavior," 789.

26. Laurette Dube, Jordan L. LeBel, and Ji Lu, "Affect Asymmetry and Comfort Food Consumption," *Physiology & Behavior* 86 (2005): 564.

27. Mark A. Newcombe, Mary B. McCarthy, James M. Cronin and Sinead N. McCarthy, "'Eat Like a Man': A Social Constructionist Analysis of the Role of Food in Men's Lives," *Appetite* 59 (2012): 391–98.

28. Harriett Bruce Moore, "The Meaning of Food," *The American Journal of Clinical Nutrition* 5 (1957): 80–81.

29. Ibid., 392.

30. Ibid., 393.

31. Ibid., 397.

32. Carolyn Korsmeyer, *Making Sense of Taste: Food and Philosophy* (Ithaca: Cornell University Press, 1999), 5.

33. Ibid., 31.

34. Ibid., 168.

35. Fabio Parasecoli, *Food and Men in Cinema: An Exploration of Gender in Blockbuster Movies*, (PhD. diss., University of Hohenheim, 2009), vii.

36. *Hannah and Her Sisters*, directed by Woody Allen (1986; Los Angeles: 20th Century Fox, 2001), DVD.

37. *The Kids Are Alright*, directed by Lisa Cholodenko (Universal City, CA: Focus Features, 2010), DVD.

38. Steve Zimmerman, *Food in the Movies*, 2nd edition (Jefferson, NC: McFarland & Company, Inc., 2010).

39. *The Upside of Anger*, directed by Mike Bender (2005; Los Angeles: New Line Home Cinema. 2010), DVD.

40. "Restrained eating refers to a persistent pattern of eating-related cognitions and behaviours in order to reduce or to maintain body weight." Macht, "How Emotions Affect Eating," 2.

41. This line is included in the trailer for the film *The Holiday*, directed by Nancy Meyers (2006; Culver City: Sony Pictures Home Entertainment, 2007), DVD.

42. Cheryl Thompson, "Neoliberalism, Soul Food, and the Weight of Black Women," *Feminist Media Studies* (2015): 4. DOI: 10.1080/14680777.2014.1003390.

43. *Something New*, directed by Sanaa Hamri (Universal City, CA: Focus Features, 2006), DVD.

44. *The Single Mom's Club*, directed by Tyler Perry (Santa Monica, CA: Lion's Gate, 2014), DVD.

45. *Shallow Hal*, directed by Bobby Farrelly and Peter Farrelly (2001; Los Angeles: 20th Century Fox, 2002), DVD.

46. *Disfigured*, directed by Glenn Gers (Canoga Park, CA: Cinema Libre, 2008), DVD.

47. *In Her Skin,* directed by Simone North (2009; Orland Park, IL: MPI Home Video, 2011), DVD.

48. *Fat Girl*, directed by Catherine Breillat (2001; New York: Criterion, 2004), DVD.

49. *Last Holiday*, directed by Wayne Wang (Los Angeles: Warner Brothers, 2006), DVD.

50. *Seeking a Friend for the End of the World*, directed by Lorene Scafaria (Universal City, CA: Focus Features, 2012), DVD.

51. *This is 40*, directed by Judd Apatow (2012; Universal City: Universal Pictures, 2013), DVD.

52. *Tammy*, directed by Ben Falcone (Burbank, CA: New Line Cinema, 2014), DVD.

53. *Sex and the City: The Movie,* directed by Michael Patrick King (Burbank, CA: New Line Home Video, 2008), DVD; *Sex and the City 2*, directed by Michael Patrick King (Burbank, CA: New Line Home Video, 2010), DVD.

54. *A Big Love Story*, directed by Ryan Sage (2012; Chatsworth, CA: Osiris Entertainment, 2003), DVD.

55. *Chef*, directed by Jon Favreau (Hollywood: Open Road Films, 2014), DVD.

56. *Today's Special*, directed by David Kaplan (2009; Niwot, CO: Flatiron Films, 2012), DVD; see also *Delivering the Goods*, directed by Matthew Bonifacio (2011; Beverley Hills, CA: GoDigital, 2012).

57. *Eat, Drink, Man, Woman*, directed by Ang Lee (1994; Beverly Hills: MGM World Films, 2002); *Tortilla Soup,* directed by Maria Ripoll (2001; Culver City: Sony Pictures Home Entertainment, 2002), DVD.

58. *Mostly Martha,* directed by Sandra Nettelbeck (2001; Hollywood: Paramount Classics, 2003), DVD; *No Reservations*, directed by Scott Hicks (2007; Burbank, CA: Warner Home Video, 2008), DVD.

59. *The Mistress of Spices*, Directed by Paul Mayeda Berges (2005; New York, NY: Weinstein Company, 2007), DVD; *Chocolat*, directed by Lasse Hallstrom (2000; Santa Monica: Miramax Lionsgate, 2011), DVD; *Woman on Top,* directed by Fina Torres (2000; Los Angeles: Fox Searchlight, 2003), DVD; *The Recipe*, directed by Anna Lee (Seoul: Film It Suda, 2010); *Like Water for Chocolate*, directed by Alfonso Arau (1992; Santa Monica, CA: Miramax Lionsgate, 2011), DVD; *Simply Irresistible*, directed by Mark Tarlov (1999; Beverley Hills, CA: Starz/Anchor Bay, 2013), DVD.

60. *Waitress,* directed by Adrienne Shelly (Los Angeles: Fox Searchlight, 2007), DVD.

Part IV

LOCAL FOOD

Chapter 11

"Food Virtue"

Can We Make Virtuous Food Choices?

Nancy E. Snow

What is it to be virtuous with respect to food choices? In this chapter, I argue that it is to exercise practical wisdom in one's food choices such that two central values, sustenance and sustainability, are preserved. Making food choices that respect and preserve these two values constitutes what I call "food flourishing," that is, achieving a state of well-being with respect to food. I develop the notions of sustenance and sustainability in the first part of the chapter. In the second, I sketch out more fully what "food virtue" means in the practical order by taking Michael Pollan's "hunter-gatherer" meal as a foil.[1] "Food virtue" as I describe it is more practical than Pollan's hunter-gatherer meal (he admits it is impractical). Yet, my story of how virtuous food choices might be made is susceptible to two arguments that I will press against it in the third section: (1) Given the present agribusiness complex in industrialized nations, many, if not most people living in these countries are not economically positioned to be able consistently to make virtuous food choices. (2) Those of us who are so positioned cannot consistently make virtuous food choices without moral remainder; that is virtuous food choices in the present context lead us into what Hursthouse calls a "tragic dilemma."[2] We are led into a tragic dilemma, I argue, because we cannot consistently promote the values of sustenance and sustainability without perpetrating economic harm upon innocent individuals. I end on a mixed note. Without concerted collective action to change agribusiness and our present food culture, the chances of making truly virtuous food choices are slim. Yet, radical hope and moral imagination could help us to combat the forces that now limit the widespread availability of nutritious food.

I. WHAT IS IT TO BE VIRTUOUS IN OUR FOOD CHOICES?: AN INITIAL SKETCH

Aristotle thought that virtues are entrenched dispositions to perceive, respond, choose, and act from motives that express certain values, such as kindness and generosity, and that these dispositions are necessary, but not sufficient, for a flourishing life.[3] I follow Aristotle in these ideas, and seek to apply them to food choices made in contemporary industrialized nations.

To make virtuous food choices is to use practical wisdom in deciding what and how to eat. Consistently with Aristotle's views on temperance, it is to avoid eating both too little and too much, to eat just the right kind of food in the right amounts in the right way and for the right reasons.[4] Being temperate with respect to food is one of the virtues partially constitutive of flourishing. In today's world, being temperate with respect to food is a virtue that cannot be understood without reference to a panoply of other virtues, nor without understanding more specifically what it means to flourish with respect to one's food choices and eating habits. I believe we can get a better grasp of what it means to be virtuous with respect to food choices by considering what I call "food flourishing," that is, the state of well-being with respect to food. This state of well-being, as I understand it, refers not only to physical health and gustatory enjoyment, but also to ethical well-being—the state of being on firm moral ground with respect to what and how one eats. Since I agree with Aristotle that immoral or vicious behavior does not conduce to flourishing, and contend that what and how one eats is an important contributing factor to a life of flourishing or *eudaimonia*, food flourishing, as brought about through virtuous choices, is an essential component of flourishing *tout court*. So what is food flourishing?

Food flourishing consists of two basic values with respect to eating and food, namely, sustenance and sustainability. I define "sustenance" as the ability to feed oneself and others nutritious, nontoxic, tasty and appealing food, and "sustainability" to refer to the fact that the food one eats has been produced in ways that sustain the environment, animal life and human food producers. "Sustenance" should be interpreted broadly to include the notions that we should be able to take pleasure in eating nutritious food, that is, we should enjoy what we eat, and we should be able to use food in valued rituals, for example, in Passover Seders or Thanksgiving dinners. It seems reasonable to fold these notions into our understanding of sustenance, since sustenance is meant to express one central way in which we can flourish with respect to food.[5] Taking pleasure in eating and using food as part of valued rituals helps to sustain us and enables us to lead flourishing lives, not only with respect to food, but also more generally. Think of how much convivial meals with friends add to our zest and enjoyment, and how the use of food in rituals helps us to commemorate events of religious or traditional significance. The value of sustenance addresses our personal relationships with food—what we choose to eat, whether and how we enjoy what we eat, and the many and varied roles food can play in our lives. "Sustainability" is broader in scope. It speaks directly to our present practices of food production, placing a moral constraint on what and how we eat so that our food choices do not cause harm, whether it be to the environment, to nonhuman animals, or to other people. When one exercises practical wisdom in one's food choices in ways that respect and preserve the values of sustenance and sustainability, one achieves (or approximates) food flourishing. Why think these two values are of central importance for food flourishing?

These values can be explored and defended through several interlocking arguments (which, given space constraints, I offer as argument sketches, rather than as fully blown arguments). The first is a type of proper function argument that justifies the value of sustenance. Since the function of eating is to sustain our bodies, we should seek to eat nutritious, nontoxic food, as that sort of food best sustains us. Additionally, taking pleasure in what and how we eat helps us to sustain ourselves. If food is tasteless or tastes bad, we will not want to eat, and thus will at least run the risk of not adequately sustaining ourselves. So we can add to the proper function argument the notion that the nutritious, nontoxic food we should eat should also be tasty and cause us to take pleasure in eating.

The idea that food should be appealing can also be defended in terms of the proper function argument. If food is unappealing, we might not want to eat it, even if it has the other properties already mentioned. I do not mean here that food must be aesthetically pleasing in the sense of being well presented, as when one dines in an upscale restaurant. What I mean is that psychologically, humans have aversions to foods that are presented in certain ways. I am not making a very deep point here, but simply noting, on the basis of experience and anecdote, some common, and perhaps, culturally based, food aversions. Some people do not care to eat foods, such as fish or roasted pig, which still has the animal's eyes or face intact. People might be repulsed by food made to look like vomit or feces, even though the food is tasty and nutritious, and they might enjoy eating it, if only they could take that first bite. The larger point here is simply that to sustain ourselves, the food we eat must appeal to us so that we want to eat it.

One might object that soldiers in wartime are sustained by rations that are unappealing and not something they want to eat or would choose to eat under different circumstances. This objection misconstrues the meaning of "sustenance" as it is used here. As previously mentioned, the conception of sustenance which is part of food flourishing is a rich and broad notion. It is not equivalent to "sustenance" in the sense of merely keeping oneself alive, though it includes that minimal notion. Sustenance as partially constitutive of food flourishing helps us to survive and *thrive* with respect to food—to have a healthy and pleasurable relationship with food. This relationship must be chronic or ongoing, such that those who achieve the value of sustenance in their food choices typically eat food that is nutritious, nontoxic, tasty and appealing. If any of these properties of food is chronically lacking, the value of sustenance, and thus food flourishing, will be diminished or jeopardized.

One final point about sustenance. One might think it otiose to include both "nutritious" and "nontoxic" in the list of properties that food should have. If food is nutritious, how can it also be toxic? Sadly, it is necessary to include both properties on the list. This is because even nutritious foods, such as apples, or drinks, such as bottled water, can be processed or packaged in ways that allow dangerous pesticides or other toxins, such as chemicals leaching into beverages from plastic containers, to enter our food supply. Perhaps the most disturbing example is processed meat, which can contain traces of the antibiotics put into animal feed in order to keep animals healthy under the unnatural conditions in which they're raised.[6] The use of these antibiotics has been linked with the development of antibiotic-resistant illnesses in people.[7] Similar concerns have been raised about milk.[8]

So much for sustenance. What about sustainability? Why add this value? Isn't sustenance, which delineates ways in which I can flourish with respect to my food choices, enough? The reason for adding sustainability is found in the general Aristotelian perspective that flourishing is incompatible with immoral behavior. Not only is it true that Aristotle thinks vicious behavior cannot lead to flourishing, it seems part of his ethical outlook that external goods gotten in immoral ways cannot be genuine goods and consequently cannot be constituents of flourishing. If my house and my wealth, for example, were obtained immorally, or if I got my children by stealing them at birth, none of them can count as either a morally neutral or benign good, since they were acquired in immoral ways. Similarly, if food is a good I need for sustenance, and a good with respect to which I can either succeed or fail to express virtue (temperance), then if it is produced or procured in vicious ways, it cannot be part of my flourishing.

It is difficult to see how food cultivated in ways that harm the environment, animal life and human producers is not immorally produced. The evidence that agribusiness food production is harmful in these ways is glaring and has been explored in great detail elsewhere.[9] Stating sustainability as a separate value to be respected and preserved in our food choices is meant to underscore the systemic nature of the harms caused by food production in Western, industrialized nations and other countries that have adopted Western production methods. It is meant to direct our moral attention to the fact that sustenance is not the only value involved in virtuous food choices. Given the global economic and social interdependence that characterizes our world today, it is not implausible to think of the moral community as being worldwide, extending to future generations, and including both animal and nonanimal life on our planet. If we think of ourselves as citizens of the cosmos (to borrow a Stoic notion), and take moral responsibility for the choices we make, we need to realize that the value of sustainability is as much a part of what it means to flourish with respect to food choices as is sustenance.[10] In other words, we do not make food choices, even those that preserve the value of sustenance, in a social vacuum. As Aristotle recognized, the nature of the society in which we live affects our choices and the kinds of people we become through them. For him, this idea is expressed in the influence the family and the *polis* have on the individual. In this global day and age, our purview must be more inclusive. Our choices and lives are affected by larger forces and can, in turn, influence them. Sustainability is meant to broaden our vision with respect to the food choices we make.

To sum up my thoughts in this section, I submit that "food flourishing" consists in achieving or approximating the values of sustenance and sustainability in one's food choices. Sustenance is justified by a type of proper function argument, and sustainability, by the argument that we cannot achieve proper functioning or flourishing in a social vacuum. Put another way, one exercises practical wisdom or *phronēsis* when one chooses to eat the right food in the right amounts at the right time in the right way and for the right reasons—in other words, one exercises temperance. Temperate food choices respect and preserve both sustenance and sustainability. How might virtuous food choices be made in the practical order?

II. VIRTUOUS FOOD CHOICES AND HUNTER-GATHERERS

We can get a sense of how virtuous food choices might be made by comparing and contrasting them with an interesting exercise in food "production" described in Michael Pollan's book, *The Omnivore's Dilemma*.[11] Pollan describes contemporary practices of food production in the United States by taking the reader through four different kinds of meals and the production and preparation of the food in each. He focuses on four kinds of meals: a fast food meal; a meal made from "mainstream" agriculturally produced ingredients; a meal made from "alternative," that is, healthy, organic, free-range ingredients; and a "hunter-gatherer" meal. Pollan adduces compelling evidence to think the first two meals fail the sustenance/sustainability test. Though the third might pass muster with respect to sustenance, there is good reason to think there is significant deception about the nature of the products routinely sold at places such as Whole Foods Market. As Pollan documents in a chapter entitled "Big Organic," not all products sold at Whole Foods Markets and similar stores are produced locally, are organic, pesticide- and chemical-free, or free range, as consumers are often led to believe by the advertisement and promotion of food sold in such places.[12] In other words, healthy, organic "alternatives" could have many of the same sustainability issues as foods produced by mainstream agribusiness, though to lesser degrees. Free-range hens are not really free range; organic fruits and vegetables are transported in refrigerated planes and trucks, thereby incurring environmental costs; pictures of happy cows frolicking in green fields on milk cartons labeled "organic" do not reflect the realities of milking processes. In other words, alternative foods are being produced and marketed in ways influenced by mainstream agribusiness, and this diminishes their claim to be produced in ways that promote sustainability. This is not true of all "alternative" food options, but, if Pollan is to be believed, it is true of many of them. For that reason, the best way to highlight what it might mean to make virtuous food choices, I think, is to compare and contrast them with Pollan's favored meal that is produced through hunting and gathering.[13]

Assembling the ingredients for the hunter-gatherer meal, as Pollan admits, involves quite a bit of impracticality. The gathering part includes two main phases—gathering vegetables and procuring seafood. Even the gathering of vegetables is somewhat risky, as Pollan chose to gather mushrooms. He enlisted the aid of people who gather mushrooms for a living (a kind of subterranean subculture), and went with them into forests to collect his haul. There are two main problems with gathering mushrooms. First, some are poisonous, and closely resemble nontoxic types; and second, one needs to hone one's visual skills to spot any kind of mushrooms in their forest environments.

Suppose that we, as food consumers, sensibly decide not to seek to harvest mushrooms. What are our options with respect to vegetable procurement? They are twofold: grow your own, or buy them at places that guarantee they meet the criteria of sustenance and sustainability. Many people do grow their own fruits and vegetables. In many places, however, such produce cannot be grown in sufficient quantities to feed a single individual, much less a family. Thus, people turn to innovative ways of purchasing vegetables; that is, they do not go to their local mainstream grocery stores to buy produce. Farmers' markets are now becoming increasingly popular, and appear regularly on the food scene. They are local and are especially ubiquitous during the spring, summer and fall. Another option is joining a local food co-op. In my area, the Riverwest Food Co-op stocks locally grown and responsibly produced healthy food, and offers a variety of recipes and sponsored events that increase food awareness.[14]

Back to Pollan. A second aspect of the hunter-gatherer meal was Pollan's diving near the rocks of San Francisco Bay, under the tutelage of an expert, to collect seafood from around and under the rocks. This, of course, is not something we all can or should try to do. Eating seafood is especially problematic. Pollan doesn't deal with this at length, if at all, but Masson devotes a hair-raising chapter to fish farming, its cruelty to fish, who are kept in crowded conditions, and its environmental hazards, such as toxic waste seeping from holding pens into larger waterways.[15] Here, too, creative options are emerging, for example, Sweet Water Organics, a Milwaukee-based urban agricultural center located in an abandoned industrial building that uses aquaponic technology to produce fish and organic vegetables. They have posted a quote from *The Wall Street Journal* on their website: "Sweet Water has converted a former crane factory in Milwaukee into an indoor wetland, raising about 80,000 fish in tanks topped by beds of lettuce and other crops."[16] Their website also proclaims:

> Sweet Water's sustainable aquaponics system was inspired by Will Allen's . . . three-tiered, bio-intensive, simulated wetland. In the re-circulating systems, the fish waste acts as natural fertilizer for plant growth and the plants act as a water filter. Our current vegetation includes various lettuce and basil, watercress, tomatoes, peppers, chard, and spinach. Our fish are tilapia and perch. As Sweet Water grows we look to our communities and restaurants to determine what comes next.[17]

The third portion of Pollan's hunter-gatherer meal is, for me, morally problematic as well as practically unrealistic. That is the phase in which he describes hunting and preparing a wild boar. Not all of us can hunt wild boar, but hunting is popular and some people I know hunt deer and eat venison. As a moral vegetarian, hunting is problematic for me, but that is another paper and other arguments. I find two things disturbing about the description of Pollan's boar hunt. First, his narrative is interlaced with comments about our evolutionary heritage. For some reason he seems to think that hunting wild boar with a shotgun is a way of reenacting scenes from our evolutionary past, in which our ancestors were forced to hunt wild animals (though not with shotguns—with lesser weapons and more ingenuity) in order to survive. I find these comments shallow and misplaced. Hunting wild animals for our meat supply is not adaptive in our circumstances, as it was for our distant ancestors. More disturbing is his description of a boar he grazed with a shot, which fled, bleeding, frightened and in pain, back into the forest. Pollan barely mentions the harm and pain he caused this innocent victim of his desire to hunt for his meal, focusing instead on his own feelings about the boar he actually killed.

As a moral vegetarian, I do not regard meat eating as morally justified. If one does, however, meats not produced by factory farming are available in co-ops, and I take it can be purchased at some farms, thereby obviating the need for hunters to risk maiming animals who are then left to suffer and fend for themselves.

I belabor this discussion of Pollan's meal to underscore that, for three important types of food—produce, fish and meat—options such as growing one's own produce, shopping at farmer's markets, food co-ops and farms themselves, and ensuring that what one buys is not factory farmed but produced through environmentally responsible technology that does minimal harm to animals—are ways in which people can make virtuous food choices. So I hope to have made an initial case that it can be done, but care and effort must be taken in doing so. Virtuous food shopping and eating are possible and far more practical than going into forests to hunt and gather. In the next section, I develop two arguments that put pressure on this conclusion.

III. BUT ARE OUR CHOICES REALLY VIRTUOUS?: TWO ARGUMENTS TO GIVE US PAUSE

Two arguments put pressure on this initial case for the possibility of virtuous food choices. The first is an argument that making virtuous food choices as described in the first two sections of this chapter falls foul of distributive justice. The second is that making virtuous food choices places us in a "tragic dilemma." Consistently with my earlier caveat, both are argument sketches.

The first goes like this. Given the present agribusiness complex in industrialized nations, many, if not most people in these nations are not economically positioned to be able consistently to make virtuous food choices. Many people, especially the poor and the elderly, live in areas called "food deserts," in which fresh, nutritious food is either not for sale or is so overpriced as to be beyond their reach. According to McLean, "A 2009 study by the US Department of Agriculture found that 2.3 million households do not have access to a car and live more than a mile from a supermarket. Much of the public health debate over rising obesity rates has turned to these 'food deserts,' where convenience store fare is more accessible—and more expensive—than healthier options farther away."[18] Food deserts are also found in Europe.[19]

Other people, though not living in food deserts, do not have the disposable income to spend on higher-priced, fresher, organically grown fruits and vegetables, nor on meats that have not been factory farmed, nor on fish that are not the produced by industrial aquaculture. Many of those who cannot afford to buy the kind of food that promotes food flourishing as earlier described, that is, that preserves the values of sustenance and sustainability, are members of minority groups whose incomes are lower than those of many whites. Studies show that the United States suffers from serious economic inequality, often dividing along racial and ethnic lines.[20]

Among the injustices to which economic inequality gives rise is lack of equal opportunity. Economic inequality denies those in lower income brackets equality of opportunity with respect to healthy, virtuous food choices. They experience injustice in the distribution of opportunities to buy good food, and thus, are unable to enjoy fully and equally with their fellow citizens the goods of food and nutrition. Systemic inequality with respect to food options actively harms those condemned to ingest cheap, readily available fast food, junk food, fruits and vegetables sprayed with pesticides, meat, poultry and fish fed on grain that has been loaded with antibiotics, and so on. Consequently, any putatively virtuous food choices that are made by those with the requisite income cannot be regarded as truly virtuous, for they occur within a larger context of systemic injustice. Additionally, to the extent that those who are economically positioned to make such putatively virtuous food choices ignore the larger system of distributive injustice that condemns others to bad food and bad food choices, the economically better-off are complicit in perpetuating an unjust and harmful system. So, when we view what initially seem to be virtuous food choices in the wider context of the system in which they are made, we can see that they are not really virtuous at all, but instead, privileges had only by those able to afford them, or, more precisely, advantages enjoyed at the expense of others.

One might respond by saying that free market forces determine the price of food. Some foods are cheaper because they cost less to produce and market, and their lower price is owing to the completely impartial, and thus, just, invisible hand of free market forces. In the case of food in the United States, this assertion is factually false. The U.S. government subsidizes large portions of agribusiness, as Pollan and others document.[21] Moreover, many industry representatives advise the USDA and its regulators, and are able through political influence to tailor industry standards to allow for the cheap mass production of food.[22] Such production practices, though profitable for agribusiness, cause terrible cruelty to fish and animals, harm the soil and aquatic environments, have put family farms out of business, and yield unhealthy food for consumers. Moreover, agribusiness seeks to hide its practices from the public eye, as a recent editorial in the *New York Times* indicates. In an editorial entitled, "Eating With Our Eyes Closed," the editors decry "ag-gag" laws passed by Iowa, Utah and Missouri, and at the time, being considered by California, Illinois and Indiana.[23] The editors write: "In most of the major agricultural states, laws have been introduced or passed that would make it illegal to gather evidence, by filming or photography, about the internal operations of factory farms where animals are being raised."[24] This, the editors point out, is not in the public interest, since food consumption gives the public an interest in how food is produced. The industry's efforts, in collusion with supportive politicians, to hide production practices from public view is an admission of guilt, and yet another form of protectionism for an industry that can in no way claim to be supported by free market forces (if such exist).

Let's now turn to the second argument that puts pressure on the possibility of making virtuous food choices, namely, that making such choices places us in what Hursthouse calls a "tragic dilemma."[25] In a tragic dilemma, one cannot make a virtuous choice without moral remainder—either one's own life or the lives of others are harmed or marred. In the case of virtuous food choices, one cannot choose virtuously for oneself without harming innocent others. When we choose to eat organic foods, nonfactory farmed fish and meats, those who are most directly harmed are not the people who cannot afford to make such choices, but are those who are forced to participate in agribusiness in order to make a living. I have in mind here farmers in Iowa, such as George Naylor, interviewed by Pollan, whose family farmed for generations.[26] He wanted to retain his land and his farm, but was forced by agribusiness pressures to corporatize his farming and production methods, introduce genetically modified corn, and, with other similarly situated farmers, contribute to changing the landscape and environment of rural Iowa. I think, too, of the farmer interviewed in the film *Food, Inc.*, who defied Monsanto and sought to plant non-genetically modified corn.[27] His legal battles against the gigantic corporation failed. I am also thinking of the many individuals employed in mainstream grocery stores, in production, harvesting, planting and so on.[28] Agribusiness is huge, and many people depend on it. When we choose not to participate in the system, we risk harming the livelihoods of these people. Should alternatives to agribusiness actually become a threat to profitability, executives of huge companies like Tyson chicken will not be hurt. Employees on the assembly line will be.[29] Our choice is indeed a tragic dilemma. If we make what we believe to be virtuous food choices for ourselves and our families, we promote laudable values and enable ourselves to flourish with respect to food. But by flouting an entrenched and powerful agribusiness system with considerable political clout in which benefits and burdens are not equitably distributed, we risk harming many individuals whose livelihood depends on the economic success of that industry.

IV. CONCLUSION: WHY SHOULD WE CARE?

In conclusion, let's ask why we should care. Why should we care if our fellow citizens are unjustly prevented from making virtuous food choices and eating nutritious food? Why should we care if our own choices might cause economic harm to those whose livelihoods depend on agribusiness—some of whom were forced by economic necessity to change time-honored farming practices, or to take employment they would otherwise not have chosen? Why should we care if animals are cruelly treated—born and kept in appalling and unhealthy conditions, force-fed grains not natural to their digestive systems, and pumped with antibiotics to prevent the diseases they inevitably would contract under such conditions, until they are taken to inhumane slaughter? Why should we care if the environment is damaged by unsustainable farming and aquaculture practices?

The answer is deceptively simple and tragically difficult. We have to care if we want to be virtuous. "Why be virtuous" is a question well beyond the scope of this short chapter. My point, however, is that virtue, though a matter of personal choice, must be viewed expansively enough to allow us to appreciate that our choices do not occur in a social or environmental vacuum. Our choices are contextualized by our social conditions. As citizens of industrialized nations who want to be virtuous and to flourish, we should worry if our virtuous choices are privileges to be had at the expense of others. We should worry if they would cause harm to innocent people entangled in an immoral system because of economic necessity. Does this mean we should "give up the ghost" and go back to McDonald's? That we should not make virtuous food choices even if we can afford to?

I would not advise this, because I think there are many good reasons not to eat the mass-produced food that is so harmful to us, animals, and the environment. However, I would repeat the somber conclusion I mentioned in the introduction of this chapter: without concerted collective action to change agribusiness and our present food culture, the chances of making truly virtuous food choices are slim.

That is a rather dark note. A brighter view is suggested in a fascinating paper on radical hope and climate change.[30] Radical hope, a notion introduced in a book of the same name by Jonathan Lear, examines the history of the Crow nation in the face of the destruction of their traditional ways of life under white domination.[31] The essence of radical hope as attributed to the Crow is the ability to carry on despite the inability to clearly envision a future. In other words, radical hope is the impetus that spurs us forward into a future whose form we cannot conceptualize. The Crow could not conceptualize traditional Crow ways of going on under white domination. To oversimplify, Thompson argues that we cannot conceptualize what the future on earth will look like, given the present pace of global environmental degradation, and even given present and increased efforts to decrease global warming.[32] Radical hope, he contends, is needed to pull us through. Perhaps radical hope gave rise to the Riverwest Co-op, Sweet Water Urban Agriculture, the appearance of local farmers' markets, and other creative ways of growing and marketing food. Perhaps radical hope is now animating fast food workers who are walking off the job to lobby for an increase in the minimum wage.[33] Radical hope in the possibility of collective action could be our best chance to create the social conditions under which we can make truly virtuous food choices.[34]

NOTES

1. See Michael Pollan, *The Omnivore's Dilemma: A Natural History of Four Meals* (New York: Penguin Books, 2006).

2. Rosalind Hursthouse, *On Virtue Ethics* (Oxford: Oxford University Press, 1999), 74–75.

3. Aristotle, *Nicomachean Ethics*, trans. Terence Irwin (Indianapolis, Indiana: Hackett Publishing Company, 1985), 44–45. Possible exceptions to this structure are found in virtues such as self-control, patience and perseverance. Actions that exhibit these virtues seem to be done from other motives, for example, one might persevere for the sake of justice, or be self-controlled or patient for the sake of kindness. Of course, one can imagine someone trying to be self-controlled or patient for the sake of building up those virtues in herself, but often this is not because she values them for their own sakes. She values these virtues because she believes they will help her to achieve other goals. Her motivation is to achieve these goals.

4. See *Nicomachean Ethics*, 79–85.

5. I am grateful to the audience at the North American Society for Social Philosophy Thirtieth Annual Conference at Quinnipiac University in Hamden, Connecticut, for pressing the points about taking pleasure in food and using food in rituals, as well as for other helpful comments. The paper was presented on July 12, 2013.

6. See Pollan, *Omnivore's Dilemma*, 78–79.

7. See Pollan, *Omnivore's Dilemma*, 78–79; "Frontline Antibiotic Debate Overview," accessed July 28, 2013, www.pbs.org/wgbh/pages/frontline/shows/meat/safe/overview.html.

8. See William Neuman, "F.D.A. and Dairy Industry Spar Over Testing of Milk," accessed July 28, 2013. www.nytimes.com/2011/01/26/business/26/milk/html.

9. See, for example, Eric Schlosser, *Fast Food Nation: The Dark Side of the All-American Meal* (New York: Houghton Mifflin, 2002); Pollan, *Omnivore's Dilemma*; Marie-Monique Robin, *The World According to Monsanto: Pollution, Corruption, and the Control of Our Food Supply*, trans. George Holoch (New York: New Press, 2009); Jeffrey Moussaieff Masson, *The Face on Your Plate: The Truth about Food* (New York: W. W. Norton & Company, Inc., 2009).

10. See Allen Thompson, "Radical Hope for Living Well in a Warmer World," *Journal of Agricultural and Environmental Ethics*, 2009. DOI: 10.1007/s10806-009-9185-2, for a similar outlook regarding climate change.

11. Pollan, *Omnivore's Dilemma.*

12. Ibid.

13. See Pollan, *Omnivore's Dilemma*, esp. ch. 18, "Hunting: The Meat," and 19, "Gathering: The Fungi."

14. See Riverwest Co-op Homepage, www.riverwestcoop.org. Accessed April 9, 2013.

15. Masson, *Face on Your Plate,* ch. 3.

16. "Street Water Homepage," www.sweetwater-organic.com. Accessed April 9, 2013.

17. Ibid.

18. Bethany McLean, "Food Deserts in America, Illustrated," 2011, accessed April 10, 2013, http://www.huffingtonpost.com/2011/01/04/food-deserts-map_n_804110.html.

19. See "Maps Page of fooddeserts.org," accessed November 20, 2014, www.fooddeserts.org/images/kontentMaps.htm. For the United States, see "United States Department of Agriculture: Food Access Research Atlas," accessed November 20, 2014, www.ers.usda.gov/data-products/food-access-research-atlas/go-to-the-atlas.aspx.

20. See, for example, Uri Berliner, "Haves and Have-Nots: Income Inequality in America," 2007, accessed April 10, 2013, http://www.npr.org/templates/story/story.php?storyId=7180618; Alan Jenkins, "Inequality, Race, and Remedy," *The American Prospect*, 2007, accessed April 11, 2013, http://prospect.org/article/inequality-race-and-remedy; Shelly J. Lundberg and Richard Startz, "Inequality and Race: Models and Policy," 1998, accessed April 10, 2013, http://www.econ.washington.edu/user/startz/Working_Papers/Inequality.PDF.

21. See Pollan, *Omnivore's Dilemma,* 62–64; Schlosser, *Fast Food Nation*, 227–29; Daniel Imhoff, "Overhauling the Farm Bill: The Real Beneficiaries of Subsidies," *The Atlantic*, 2012, www.theatlantic.com/health/archive/2012/03/overhauling-the-farm-bill-the-real-beneficiaries-of-subsidies/254422/.

22. See Schlosser, *Fast Food Nation*, ch. 9.

23. See Editors, "Eating With Our Eyes Closed," *The New York Times*, April 10, 2012, A20.

24. Ibid.

25. Hursthouse, *On Virtue Ethics,* 74–75.

26. Pollan, *Omnivore's Dilemma,* ch. 2: "The Farm."

27. *Food, Inc.*, directed by Robert Kenner, (Participant Media. 2008. Film).

28. See Schlosser, *Fast Food Nation*, chs. 3, 7 and 8, for working conditions in the fast food industry, the effects of agribusiness on communities, and conditions in the meat packing industry.

29. But see Mark Bittman,"Fast Food, Low Pay," *The New York Times*, July 26, 2013, A19, for a report on fast food workers who are walking off the job to demand pay raises and are having some success.

30. Thompson, "Radical Hope."

31. Jonathan Lear, *Radical Hope: Ethics in the Face of Cultural Devastation* (Cambridge, MA: Harvard University Press, 2006).

32. Thompson, "Radical Hope."

33. See Bittman, "Fast Food, Low Pay."

34. Thanks to Jill Dieterle for inviting me to contribute to this volume.

Chapter 12

Limits on Locavorism

Liz Goodnick

Food justice advocates seek to absolve inequities in our food system through promoting food security (ensuring that all people have access to sufficient, nutritious food) and supporting food sovereignty (championing the right people have to define and control their own food system).[1] Thus, many proponents of food justice see locavorism as the answer.[2] The movement—hence its name—focuses on acquiring food (and, to a lesser extent, other goods) from a local source, and thus limiting "food miles." It has many benefits according to food justice advocates: it puts individuals in control of their food supply, it alleviates the obvious costs of transporting food, it reduces use of and dependence on fossil fuels, and it decreases dependence on an industrialized food chain.

But focusing solely on the "local" doesn't fully capture the heart of the movement. The benefits of sourcing food locally are overridden if, for example, the food is raised on a "Concentrated Animal Feeding Operation" (known as a CAFO[3]). While it might be true that buying meat from a local CAFO saves food miles, and thus reduces some use of fossil fuels and transportation costs, these benefits are outweighed by (1) the fact that individuals are beholden to large corporations for their food supply and (2) the overwhelming harms to the environment that industrial farms are known to cause. Since one of the primary motivations for adopting a locavore's lifestyle is to minimize environmental costs, reducing food miles is only one part of the locavore's overall aim—further aims include various practices devoted to sustainability and minimizing environmental damage.

There is a trend in the food justice movement to encourage not only the purchase of locally raised meat, but also of meat that is more sustainably raised than it is on CAFOs. In fact, several proponents of locavorism argue that eating meat is necessary to eat a truly just, local and sustainable diet.

In this chapter, I argue that while the benefits of locavorism are important, locavores fail to attend to the significant value that nonhuman animals have. Drawing on the work of Tom Regan and Richard Swinburne, I argue that nonhuman animals have inherent value. I support a no-meat eating principle derived from a principle of respect for nonhuman animals, claiming that insofar as nonhuman animals have inherent value, eating them is unjust and therefore not permitted. I conclude that the locavore's argument in favor of meat eating (which relies on a premise that nonhuman animals have only instrumental value) is unsound. In section IV, I consider two possible alternative premises the locavore might use to support ethical meat eating: that nonhuman animals have less inherent value than the environment; and that nonhuman animals suffer less by being raised for meat than the environment would suffer if we did not raise animals for meat. I argue that both of these premises are false, especially given alternative means of acquiring locally grown plant-based food. In the fifth section, I consider some implications of my argument, claiming that while my argument does require a shift in our relationship with nonhuman animals, it does not result in a complete rejection of using animals for means (so long as they are not used solely as a means). Because my argument does not have absurd consequences, I conclude that it is not permissible to eat meat in order to eat a local diet.

I. LOCAVORISM AND EATING MEAT: THE ARGUMENT

Some locavores argue that to genuinely respect nature—to eat a sustainable diet—it is not just permissible, but necessary to eat meat (depending on one's location). No locavore argues that meat should be produced in a way that is unsustainable. Instead, there is a trend in the locavore movement to encourage the purchase of "happy," "humane," "compassionate," or "ethical" meat. The idea is that eating a vegan (and probably also a vegetarian) diet is morally worse, in certain locations, than eating meat. For example, Pollan claims:

> The vegan utopia would also condemn people in many parts of the country to importing all their food from distant places. . . . The world is full of places where the best, if not the only, way to obtain food from the land is by grazing (and hunting) animals on it—especially ruminants, which alone can transform grass into protein. To give up eating animals is to give up on these places as human habitat, unless of course we are willing to make complete our dependence on a highly industrialized national food chain. That food chain would be in turn even more dependent than it already is on fossil fuels and chemical fertilizer, since food would need to travel even farther and fertility—in the form of manures—would be in short supply. Indeed, it is doubtful you can build a

> genuinely sustainable agriculture without animals to cycle nutrients and support local food production.[4]

While Pollan states that eating meat is only necessary in certain locations (but these locations are numerous—so numerous that it may be impossible given the current human population to avoid them), he implies that eating meat may be necessary in any location. Other locavores argue that it is not possible to grow food anywhere in a sustainable way without using nonhuman animals. For example, Kathy Rudy claims, "From an environmental perspective, removing animals from farms has been devastating. There has never once been a healthy ecosystem on the planet that did not include animals. There has never once in all of human history been a culture that farmed without any animals."[5] Rudy's ideal is that all people ought to buy their food from small, local, pastoral farms:

> The current revival of the small farm . . . practices the kind of sustainable agriculture that has been in existence on the planet for over 12,000 years. Plants and animals occupy the same space at different times; mammals such as pigs, goats, and sheep eat the parts of the plant that we cannot eat (stalks, leaves, vines, etc.), along with many weeds. Poultry thrives on bugs and pests. Ruminants such as cattle and buffalo graze nonarable land and transform something we cannot eat (grass) into something we can (meat). These animals all excrete a fertilizer that not only nourishes future plants, it also anchors the topsoil and keeps it from running off. The best small farms strive to be "closed system"—meaning they import almost nothing onto the farm (animals reproduce themselves through good husbandry practices, seeds are saved from the strongest plants for next year's crops, etc.). Such self sufficiency forms the core of agricultural sustainability.[6]

Arguments concluding that locavores must eat meat either tacitly or explicitly rely on a consequentialist premise that, when making food choices, what is of the utmost importance is the overall environmental impact of one's diet. While this principle has it's roots in Aldo Leopold's "land ethic,"[7] especially as understood by J. Baird Callicott,[8] I wish to provide a principle that better incorporates the modern locavore's insistence on sustainability—the:

> **Principle of Sustainability** (POS; With Respect to Food Choices): Food choices are (morally) right insofar as they tend to maintain or enhance the quality of the environment.

Pollan endorses this principle: "If our concern is for the health of nature—rather than, say, the internal consistency of our moral code or the condition of our souls—then eating animals may sometimes be the most ethical thing to do."[9] This principle is (a) consequentialist insofar as it focuses solely on the impact of our food choices and (b) environment-centric insofar as it places the primary locus of value on the health (quality) of the environment. While not explicit, it is likely that most proponents ultimately value the environment of the entire planet; however, most of the decisions we make will directly impact our local environment, so local considerations play a central role in evaluating our food choices.

It is important to note that the argument that eating meat is required (or the argument that it is permitted) by the POS is contentious. Some (such as Henning,[10] McWilliams,[11] Singer and Mason,[12] and Stanescu,[13] just to name a few) argue that meat production, regardless of scale, is worse for the environment than growing plants. Others argue, on alternative consequentialist grounds, that eating meat is problematic—while it may be the best practice for the environment, it violates other consequentialist principles insofar as it encourages patriarchy and domination (see, for example, Pilgrim[14] or Stanescu[15]). Moreover, most locavores agree that because of the POS, meat eating must be reduced insofar as it uses a great deal of land, water and fossil fuels. For the purposes of this chapter, I will ignore these issues. Instead, I will focus on a deontological claim against meat eating.

II. LOCAVORES AND THE POS: NONHUMAN ANIMALS HAVE ONLY INSTRUMENTAL VALUE

As stated, the POS places the environment as the highest good. But just because a principle claims that the environment is the highest good doesn't entail that all else has merely instrumental good. It might be the case that while the POS should primarily govern our food choices, there are other ethical principles that would override it in the case of conflicts. For example, Sterba argues in favor of "A Principle of Human Preservation: Actions that are necessary for meeting one's basic needs or the basic needs of other human beings are permissible even when they require aggressing against the basic needs of individual animals and plants or even of whole species or ecosystems."[16] This principle entails that humans have more than instrumental value. In cases when this principle is in conflict with the POS (e.g., where the population is too big to survive without causing environmental damage), one is permitted to damage the environment with a certain food choice in order to stay alive.

While Sterba's principle is not explicitly consequentialist, it could be interpreted as such. However, it may be the case that the POS could be overridden by a more obviously deontological principle premised on the inherent value of other beings. For example, suppose it turned out that killing humans for food, in some circumstances, tended to enhance the quality of the environment. Many people would reject this practice, supporting a ban on cannibalism and a principle such as:

> **No-Cannibalism Principle (NCP):** It is always wrong to kill humans for food.

When faced with a conflict, people would choose to violate the POS in favor of the NCP because humans have inherent value.[17] Thus, the NCP is supported by a more general principle:

> **Principle of Respect for Humans (PRH):** We are to treat humans in ways that respect their value.[18]

Because humans have inherent value, we must respect them. Killing and eating humans does not respect their value insofar as it treats them as a mere means. Thus, it is always wrong to kill humans for food.[19] Based on the PRH, no human can be killed and eaten regardless of the potential good consequences of doing so.

In endorsing the NCP, one claims that humans have inherent value. But insofar as the POS is used in arguments for meat eating, with no corresponding PRA, its proponents implicitly claim that nonhuman animals have no inherent value (and merely instrumental value).

Rudy, however, argues against this point. She claims that locavores do value nonhuman animals:

> The vast majority of farmers in locavorism have a reciprocal and connected relationship with their animals; they name them, they provide care for them, they allow them room to roam outside, they let them live much longer lives than industrial farms, they encourage mothers to care for their young, they feed them well, provide clean and warm shelter (sometimes in the human's home), they often spare those special animals that seek human connection, and mostly they dread slaughter.[20]

While it might be true that the farmers described don't *see* their nonhuman animals as having no inherent value, it is not clear that the idea of owning nonhuman animals is consistent with their having inherent value. As Stanescu writes:

> What most concerns me is that these expressions of feelings of care for animals serve to mask the simple reality that for the entirety of their lives, these animals live as only buyable and sellable commodities, who exist wholly at the whim of their "owners." As long as animals can be owned, bought, sold, and treated at their "owners" whims, the concept of loving animals will have little impact.

> While some individual "owners" may choose to treat their animals better than others, the idea that animals can be "owned" at all reinforces the notion that animals exist only as human property and, as such, the concept of loving animals loses its power as a societal critique.[21]

It's possible that meat eaters accept that nonhuman animals have inherent value (as I will explain in section IV), but I am sympathetic to Stanescu's argument. Even if a meat eater doesn't assume nonhuman animals have merely instrumental value, she thinks that their value is such that they can be used for the purpose of human nutrition. So if nonhuman animals have any inherent value, it is easily overridden. Hence, I will proceed in the next section as though meat-eating locavores assume that nonhuman animals have only instrumental value, since even if this is strictly speaking false, it's true insofar as nonhuman animals have no value that prevents their use by humans as food.

III. IN FAVOR OF A PRINCIPLE OF RESPECT FOR NONHUMAN ANIMALS

So far, I have claimed that the locavore who advocates for sustainable meat eating relies on the POS. I have then argued that locavores who use the POS to support ethical meat eating believe that individual nonhuman animals have no inherent value. According to the locavore, we cannot make deontological claims against eating meat as we might in the case against cannibalism. In what follows, I will argue that this view is incorrect. I argue that nonhuman animals do have inherent value, and in favor of

> **Principle of Respect for Nonhuman Animals (PRA):** We are to treat nonhuman animals in ways that respect their value.

This entails a parallel to the NCP, the

> **No-Meat Eating Principle (NMP):** It is wrong to kill and eat nonhuman animals for food.

Because animals have inherent value, we must respect them. Killing and eating animals does not respect their inherent value insofar as it treats them as a mere means. Thus, it is always wrong to kill animals for food. Therefore, locavores are not permitted to eat meat insofar as they base their decision on the POS. The NMP overrides the POS just as the NCP overrides the POS, forbidding us to kill and eat humans even when doing so would have the best consequences. While there may be costs of not eating meat, these costs are trumped by the duty to refrain from killing nonhuman animals for meat.

At this point in my argument, one might object that having inherent value is not a sufficient condition guaranteeing a corresponding right to not be killed and eaten for food (in some wide range of circumstances)[22] since it's not the case that having respect for humans is the reason why we don't eat them. One might argue that the PRH isn't the grounds for the NCP; thus a corresponding PRA doesn't entail the NMP. Following Cora Diamond,[23] one might argue that we oppose cannibalism for some other reason besides a generic respect for humans—one that has to do with our relationships. For example, we must live in such a way that we do not relate to each other as possible sources of food in order to live a flourishing life. Another possibility is that we incur special obligations to humans in virtue of our capacities to engage one another cooperatively. So, one might argue, since we don't have these same relationships with nonhuman animals traditionally used as food, we don't have a similar obligation toward them.

It seems that what Diamond has done is to point out additional sufficient reasons to resist eating humans. However, I only claim that the PRH is sufficient for the NCP. Pointing out additional sufficient conditions for an exemption from being eaten doesn't entail that the PRH is insufficient. The fact that we have special obligations to humans in virtue of our relationships with them probably does give us additional reasons not to eat them—reasons that may be relevant in cases of conflict between human and nonhuman animal lives or between certain animals (like our pets) and other animals (like "vermin"). But this does not mean that their inherent value isn't a reason not to eat nonhuman animals.

Still, if the objector instead suggests that having inherent value isn't sufficient for exempting one from being consumed as food (in a wide range of usual circumstances), and that the reasons suggested are not additional but alternative ones against cannibalism, I disagree. It seems obvious that using something as a mere means to our ends does not respect its inherent value.[24] And killing something for food seems to be a paradigm example of using something as a mere means. If you didn't know the moral status of a creature, the default position would be to refrain from killing and eating it. Thus, the burden of proof is on the opposition—because these statements are intuitively true, I demand a much more convincing positive argument against them in order to reject them, and offering *additional* sufficient conditions not to eat humans is not a convincing argument.

Of course, for my argument to work, it must be the case that nonhuman animals do deserve our respect. I will argue in favor of the PRA (and therefore the NMP) by drawing on the work of Tom Regan[25] and Richard Swinburne.[26]

In *The Case for Animal Rights*, Regan endorses a general Principle of Respect:

> **Regan's Respect Principle (RRP):** "We are to treat those individuals who have inherent value in ways that respect their value."[27]

He goes on to argue that (many) nonhuman animals have inherent value, so we ought to respect that value. Regan claims that a sufficient (but not necessary) condition for an individual to have inherent value is that the individual is the subject of a life. He explains:

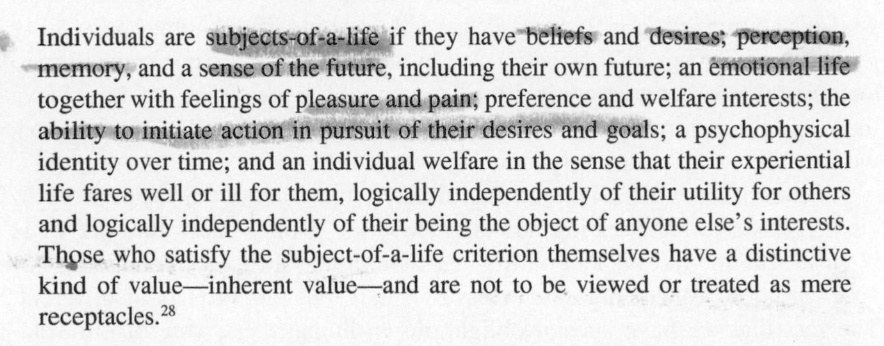

> Individuals are subjects-of-a-life if they have beliefs and desires; perception, memory, and a sense of the future, including their own future; an emotional life together with feelings of pleasure and pain; preference and welfare interests; the ability to initiate action in pursuit of their desires and goals; a psychophysical identity over time; and an individual welfare in the sense that their experiential life fares well or ill for them, logically independently of their utility for others and logically independently of their being the object of anyone else's interests. Those who satisfy the subject-of-a-life criterion themselves have a distinctive kind of value—inherent value—and are not to be viewed or treated as mere receptacles.[28]

One way to use Regan's criterion (though not exactly the way he uses it) is to argue that he has explained what it is about humans, including those who do not have the capacity for moral action, that gives them inherent value. We can then argue (as Regan does), that many nonhuman animals are the subject of a life. Since what happens to the subject of a life matters to the subject, we have moral duties with respect to it (even if it is not a moral agent, but a moral patient). Together, RRP and the claim that nonhuman animals have inherent value entail the PRA.

Regan's claim that "given the postulate of inherent value, no harm done to *any* moral agent can possibly be justified merely on the grounds of its producing the best consequences to all affected by the outcome"[29] supports the NCP. One is not permitted to cannibalize, even if the consequences would be spectacular, because humans have inherent value. Likewise, if Regan is correct and nonhuman animals also have inherent value, then we can justify the NMP in the same way.

However, it's not clear *why* being the subject of a life is sufficient for having inherent value. Others have given alternative criteria. For example, following Peter Singer[30] (but with a decidedly deontological twist), one could argue that sentience alone is a sufficient condition. Following a tradition in Christian thought, Swinburne[31] argues that nonhuman animals have inherent value insofar as they intentionally and spontaneously act with purpose, and are unconflicted and unopposed (though not free in a robust sense). He claims that such actions are intrinsically good, and because some nonhuman animals are capable of acting in this way, they have inherent worth. For Swinburne, the ability to exercise causal power in a self-directed way is sufficient for inherent value.

I suggest that we approach this problem in a slightly different way, by thinking about a new invention—a kind of experience machine for chickens called "Virtual Free Range." While it might not make clear the sufficient conditions for inherent value, this reality-based thought experiment provides evidence that many nonhuman animals do *have* inherent value. Regan points out that there is an important difference between those individuals with inherent value and those without. Individuals with inherent value are more than just receptacles for pleasurable experiences. He explains, "The inherent value of individual moral agents is to be understood as being conceptually distinct from the intrinsic value that attaches to the experiences they have (e.g., their pleasure or preference satisfactions), as not being reducible to values of this latter kind, and as being incommensurate with these values."[32]

Nozick shows (among other things) that humans have inherent value by appealing to the experience machine (a kind of virtual reality in which you are hooked up to some machine in a lab, but have the experiences as though you were living in the "real world").[33] He claims that we should not want to enter the experience machine, even if in the experience machine our lives would be fully pleasurable, or fulfilling, or the like. Although he gives several reasons to reject the experience machine, lack of freedom and access to the truth are among them. Even if we cannot point to exactly what is missing in the experience machine, the fact that we wouldn't enter it suggests that we are not mere receptacles; we are valuable for something besides our experiences; we have inherent worth.

Nozick admits to not being precisely sure what it is that matters besides our experiences. But, if we have the same intuitions regarding nonhuman animals, and if we can neither elaborate what it is that makes the experience machine for humans problematic nor conclude that this does not apply to nonhuman animals, then we can also conclude that those nonhuman animals have inherent value. Nozick explains:

> Without elaborating on the implications of this, which I believe connect surprisingly with issues about free will and causal accounts of knowledge, we need merely note the intricacy of the question of what matters *for people* other then their experiences. Until one finds a satisfactory answer, and determines that this answer does not *also* apply to animals, one cannot reasonably claim that only the felt experiences of animals limit what we may do to them.[34]

In 1974, the experience machine was a matter of science fiction. Today, however, things like Oculus Rift (a virtual reality headset for gaming) are headed toward making Nozick's experience machine a reality. Recently, a company called Second Livestock began work on an Oculus Rift-type virtual reality headset for chickens. As described on the company website, "Eliminating the need for the physical space required for free-range livestock our Virtual Free Range™ gives livestock the experience of Free Range life while in the safe confines of our facility."[35] In short, the company is working on creating an experience machine for chickens. If it's true that nonhuman animals are just receptacles of experiences, and have no inherent value, then keeping the chickens in the conditions similar to a CAFO while outfitting them with the Virtual Free Range device is perfectly acceptable.

There are, however, reasons to resist this claim. While the chickens in the experience machine (presumably) would have pleasurable experiences, their lives would be worse than if they were free. While they would have pleasurable experiences, they would be entirely unfree to exercise their causal power in a self-directed way. They wouldn't have any opportunities to actually do the kinds of things that chickens like to do—pluck bugs from the grass, protect their chicks, etc. They wouldn't even get to stretch their wings or walk around. They would be in CAFOs—they just wouldn't know it. This suggests that something like being self-directed is important, even for chickens. (Though, perhaps sentience or being the subject of a life makes self-direction important.) The intuition that the experience machine for chickens is not, all things considered, the best (of course, it may be better than living in a CAFO without the technology) indicates that nonhuman animals, including chickens, do in fact have inherent value.[36]

I have argued that some nonhuman animals have inherent value, but I have not provided a list of nonhuman animals that "pass the test." Even if one does determine the necessary and sufficient conditions for having inherent value, it will still be an empirical question, requiring experts (veterinarians, animal behaviorists, neurologists, etc.) to determine which species (or even individual animals) meet the requirements. But more importantly, I think, like Nozick, that it is difficult (perhaps impossible) to provide a clear list of necessary and sufficient conditions for inherent value.

Instead, I have provided a test: if we think that a nonhuman animal's life would be worse in an experience machine, then we have reason to believe that the nonhuman animal has inherent value. I have also provided some things to consider when thinking about the Virtual Free Range machine: most importantly, self-direction. Self-direction (not necessarily robust libertarian free will) is objectively valuable since it is a necessary condition of certain virtues or actions that we think of as valuable—those that nonhuman animals can and do exercise—such as cooperation, protecting one's young, learning, altruism, sacrifice, etc. Any animal that has the capacity for self-direction will likely be one that we will have difficulty placing into the experience machine. In order to have inherent value, it is likely that nonhuman animals don't need to have a robust free will or self-consciousness (as perhaps some primates do). However, we might require more than just capacity for motion. After careful philosophical and empirical study, most common food animals will almost certainly pass the test and will be determined to have inherent value. But there will likely be reasonable disagreement and borderline cases, which may even require the rethinking of our treatment of nonanimals (plants, molds, or even bacteria).

Whatever nonhuman animals pass the test have inherent value; thus they are subjects of RRP. Thus, we ought to expand the NCP to count for nonhuman animals and endorse the NMP as well.

IV. WHEN PRINCIPLES CONFLICT

So far, I have defended the view that nonhuman animals have inherent value and have used this claim, in conjunction with a Respect Principle, to support a NMP. I will conclude by addressing two parallel areas of potential conflict. First, I will address the claim (A) that the NMP and the PRH are in conflict—that in order to survive, humans need to eat meat. Second, I will address the argument (B) that the environment has inherent value, that this generates a parallel Principle of Respect for the Environment (PRE), and that the NMP conflicts with it. This strategy could be used by a locavore who claims that it is untrue that she thinks of nonhuman animals as mere instruments, but chooses to eat meat anyway.

(A) One might claim that the NMP and the PRH are in conflict. The PRH instructs us to treat humans in ways that respect their value. This surely entails allowing them to survive. Thus, if humans require meat in order to survive, we must allow humans to violate the NMP. While some may attempt to make this argument for all humans, they are unlikely to be successful. This is an empirical question, so I will note that "it is the position of the American Dietetic Association and Dietitians of Canada that appropriately planned vegetarian diets are healthful, nutritionally adequate, and provide health benefits in the prevention and treatment of certain diseases."[37] Given that it is unlikely that most adults must eat meat to survive, this conflict is moot. However, this may not be true for pregnant women, other nutritionally vulnerable people, and the like.[38] Thus, not in all cases, but in certain cases, it may be the case that there is a legitimate conflict between NMP and PRH.

(B) One might sensibly claim that the environment itself has inherent value. Based on RRP, then, we can generate:

> **Principle of Respect for the Environment (PRE):** We are to treat the environment in ways that respect its value.

This entails that in certain cases (perhaps the ones that locavores like Pollan argue require meat eating) the PRE conflicts with the NMP.

Suppose that in both of these cases, the objector makes the decision to eat meat. This decision must be justified in light of the conflict with NMP. There are two ways in which the objector may do this. (1) The locavore could argue that while (A) both nonhuman animals and humans have value or (B) both nonhuman animals and the environment have value, the value of (A) humans or (B) the environment is higher than that of nonhuman animals. (2) Alternatively, the locavore could apply something like Regan's:

> **Worse-off Principle (WOP):** "Special considerations aside, when we must decide to override the rights of the many or the rights of the few who are innocent, and when the harm faced by the few would make them worse-off than any of the many would be if any other option were chosen, then we ought to override the rights of the many."[39]

Applying this principle, along with the claim that (A) humans or (B) the environment would be made worse-off, justifies meat eating.

Case A is significantly different from case B. In some rare cases (desert island scenarios, nutritional deficiencies, etc.), where the survival of a human is at stake, eating meat is permissible because of WOP. However, I do not think that this is the case for the locavore—either when the justification relies on the claim that the environment has more value than individual nonhuman animals, or when the justification relies on the claim that the environment would be made worse off. I'll consider each claim in turn.

First, it is false that the environment has more value than individual nonhuman animals. For one, I am sympathetic to Regan's claim that inherent value doesn't come in degrees and is thus incommensurable.[40] One might argue, however, that "lifeboat cases" show that there are levels of inherent value: if one were forced to choose between ejecting a lobster, a chicken and a monkey from a lifeboat, one should dump each in that order. But these types of cases don't show that there are degrees of inherent value. The animals all have a right to live and any death is a tragedy. But, since life isn't fair, conflicts arise. Instead of saying that the lobster is less valuable, we can account for our actions (or our intuitions about what we should do) by appeal to a principle on how to properly deal with conflicts—conflicts between equals in one respect (inherent value)—the WOP. It is likely that if we were to choose the monkey over the lobster to eject from the lifeboat, she would be worse-off than the lobster would have been if ejected not because the monkey has more inherent value, but because it has higher cognition, a more robust sense of its future, the capability to suffer more, future projects that will be destroyed, etc.

Even if we allow for degrees of inherent value, the claim that the environment has more value than individual nonhuman animals is false. Imagine two kinds of worlds: one with a healthy natural environment that includes nonhuman animals but no humans and one that has no animals whatsoever. The first world has more value than the second (though this doesn't entail that the world with no nonhuman animals has no value at all). Consider self-direction—one of the conditions connected to inherent value and possessed by many nonhuman animals. One of the reasons that being self-directed is especially relevant to value is because it allows individuals to purposefully complete certain valuable actions, such as protecting one's young and kin, sacrificing one's life for another's, and other acts of altruism. This is part of the reason that individuals who possess self-direction add value to a world, and so a world with those actions is more valuable than one without. While a world with no animals is surely valuable, it is missing the kinds of acts that require self-directed individuals.

Second, while it's possible that the environment would be harmed in Pollan's "vegan utopia," it is unclear that this is necessarily the case. This again is an empirical question. It may be true that there have *as of yet* been no sustainable farms that don't include nonhuman animals, but this doesn't require that we raise them for food. New technologies and strategies such as hydroponics, rooftop gardens, community gardening, canning/freezing operations, vertical farming etc. may provide a way to grow vegetarian food sustainably and without the use of nonhuman animals. Even if the environment is harmed, it is unlikely that, given a requirement to use whatever means possible to minimize those harms,[41] it will be made worse off than many nonhuman animals would be if meat eating were permitted, precisely because the environment is neither self-directed nor sentient nor the subject-of-a-life.

It is likely that currently, given consumption and the lack of existing technologies/infrastructure, many people are not able to survive without relying on nonhuman animal farming or hunting. It is also possible that some communities living in extreme environments (such as the Inuits in Canada or Greenland) may never be able to achieve this ideal. In these cases, we must appeal to the WOP. It's possible that in these circumstances, people will have to continue eating meat to survive; this situation, however, should be thought of as temporary and most resources should be directed toward relieving the situation in the future and toward, for example, the development of alternative, sustainable vegetable-growing and preserving technologies.

V. IMPLICATIONS

One of the most pressing issues remaining is the implications of my argument. I have argued in favor of the PRA, claiming that since nonhuman animals have inherent value, we must respect their value. One of the ways I argued we must do this is by refraining from killing them for meat. My argument that the NMP follows from the PRA relies on the premise that it is wrong to use something with inherent value as a mere means. One might sensibly wonder what the broader implications of the PRA are. Can we use nonhuman animal products as food? Can we continue having pets? Can we continue using animals for labor or for other services (e.g., to plow our fields or as service dogs)? Do any of these activities count as using animals as a mere means? If it turns out that my argument has the consequence of disallowing most interactions with nonhuman animals, it may be the case that the PRA is too strong and can rightly be rejected by the locavore.

My argument does entail that humans significantly alter their interaction with nonhuman animals. This doesn't mean, however, that its conclusions are absurd, nor does it imply that we mustn't use animals as a means. Instead, it entails that we cannot use nonhuman animals as a *mere* means. The PRA does not demand that there are no circumstances under which we can eat any animal products, keep pets, or use animals for labor or services. For one, there are corresponding principles of respect generated by the inherent value of humans, the environment as a whole, certain features of the environment (e.g., a particular forest or river), and other living things. This guarantees that there will be conflicts and so we will need to appeal to other principles, such as the WOP to determine what to do in those cases. Likely, cases will arise where using animals for food or for other things may be the best way to handle conflict. Of course, this will be highly contingent on the circumstances and the details of the kinds of lives the nonhuman animals will have. One worry is the broader consequences of these practices—much of the dairy and egg industry involves killing unwanted offspring, especially males. If this is a necessary consequence, these practices will require many claims of need from humans or the environment to offset their costs.

That said, it may be quite possible to coexist with nonhuman animals in ways that do not use them exclusively as means. For example, while service dogs do provide many valuable services for humans, their humans provide them with much in return. It may be possible to coordinate our activities with nonhuman animals in a mutually acceptable way, where we all can pursue our self-chosen projects. If this is possible, I see no reason why we cannot keep pets, use animals for labor, or even eat some animal products (such as eggs). For surely we can and often do use humans as a means in a way that respects their inherent value.

VI. CONCLUSION

The food justice movement sees locavorism as the answer to injustices in our food system: it promotes food sovereignty insofar as local communities have more power over their food supply and it supports food security insofar as it promotes ecologically sound and sustainable methods of food production. But the locavore who argues from the POS to allowing (or requiring) meat eating either assumes that nonhuman animals have merely instrumental value or assumes that their value is such that it can be easily overridden by environmental concerns. I have argued that both of these disjuncts are incorrect. First, I argued that nonhuman animals do have inherent value, and based on a Principle of Respect, I generated a NMP. If eating meat is permitted (or required) by locavorism, then the advocate of food justice should reject the movement (or at least the meat-eating portion of it) insofar as it does not entail justice for all. I went on to argue that the NMP cannot be overridden by claiming either that the environment has more inherent value than nonhuman animals or that the environment would suffer more by our refraining to eat meat than nonhuman animals would suffer by our doing so. The locavore cannot argue in favor of meat eating on the basis of the POS. I argued that while my claims do have serious implications for our current relationships with nonhuman animals, they do not require the absurd consequence that we give up all of our useful relationships with them. Therefore, it is not permissible to eat meat in order to eat a local, sustainable diet.

NOTES

1. I first presented this chapter at the 2014 ISEE 11th Annual Meeting. Thanks to all who gave comments and feedback there, especially Matt Ferkany, Beth Seacord, Sergio Gallegos and Sharisse Kanet. I presented a later draft of this chapter to the Philosophy Club at MSU Denver, and would like to thank the students for their helpful comments and suggestions. I'd also like to thank Lori Watson and Alex Hughes for their conversations, suggestions, and comments on earlier drafts. Finally, thanks to Jill Dieterle for her valuable comments on the abstract, the chapter and for putting this book together.

2. Just Food (NYC), the Detroit Food Justice Task Force and La Via Campesina are just three (among several) prominent examples. See http://www.detroitfoodjustice.

org/, http://www.justfood.org/, and http://viacampesina.org/en/ for more information. Locavorism has also garnered a great deal of support both from popular writers and activists such as Catherine Friend, Barbara Kingsolver, Michael Pollan, and Joel Salatin (of Polyface Farm), but also from academics in various fields. See Catherine Friend, *Hit by a Farm: How I Learned to Stop Worrying and Love the Barn* (Cambridge, MA: Da Capo Press, 2006); Barbara Kingsolver, *Animal, Vegetable, Miracle: A Year of Food Life* (New York: HarperCollins, 2007); Michael Pollan, *The Omnivore's Dilemma: A Natural History of Four Meals* (New York: The Penguin Press, 2006); Joel Salatin, *Folks, This Ain't Normal: A Farmer's Advice for Happier Hens, Healthier People, and a Better World* (New York: Hachette Book Group, Inc., 2011).

3. A CAFO is sometimes known as a "factory farm."

4. Pollan, *Omnivore's Dilemma*, 326–27.

5. Kathy Rudy, "Locavores, Feminism, and the Question of Meat," *The Journal of American Culture* 35 (2012): 28.

6. Ibid., 27–28.

7. Aldo Leopold, *A Sand County Almanac and Sketches Here and There* (Oxford: Oxford University Press, 1949).

8. J. Baird Callicott, "Animal Liberation: A Triangular Affair," *Environmental Ethics* 2 (1980): 311–28.

9. Pollan, *Omnivore's Dilemma,* 326–27. While the claim is a conditional, it's clear from context that Pollan endorses the antecedent.

10. Brian G. Henning, "Standing in Livestock's 'Long Shadow': The Ethics of Eating Meat on a Small Planet," *Ethics and the Environment* 16 (2011): 63–93.

11. James E. McWilliams, *Just Food: Where Locavores Get it Wrong and How We Can Truly Eat Responsibly* (New York: Little, Brown and Company, 2009). And also see James E. McWilliams, "The Myth of Sustainable Meat," *New York Times*, April 12, 2012.

12. Peter Singer and Jim Mason, *The Way We Eat: Why Our Food Choices Matter* (USA: Rodale, Inc., 2006).

13. Vasile Stanescu, "'Green' Eggs and Ham? The Myth of Sustainable Meat and the Danger of the Local," *Journal for Critical Animal Studies* 8 (2010): 8–32.

14. Karyn Pilgrim, "'Happy Cows,' 'Happy Beef': A Critique of the Rationales for Ethical Meat," *Environmental Humanities* 3 (2013): 111–27.

15. Vasile Stanescu, "'Green' Eggs and Ham?"

16 James P. Sterba, "From Biocentric Individualism to Biocentric Pluralism," *Environmental Ethics* 17 (1995): 196.

17. For this to work, the proponent of the NCP (in cases of conflict) would have to argue either that humans have more inherent value than nonhuman animals (a claim that I dispute) or that humans would be made worse off if eaten than nonhuman animals would if eaten. I discuss cases of conflict in more detail in section IV.

18. This principle closely follows RRP, but restricts it to humans only (Regan 2004, 208). I will return to a discussion of Regan's actual principle in section III.

19. One may argue that the PRH isn't the actual grounds for the NCP. Thus, when I argue in favor of a corresponding PRA, this doesn't entail a NMP. I will address this objection in the next section.

20. Rudy, "Locavores, Feminism, and the Question of Meat," 30.

21. Stanescu, "'Green' Eggs and Ham?" 108.

22. I add the parenthetical comment because, as I'll explain in section IV, the right to not be eaten can be overridden in certain (fairly extreme) circumstances. This is obvious even in the case of humans (e.g., in situations described in Piers Paul Read, *Alive: The Story of the Andes Survivors* [J. B. Lippincott, 1974]).

23. Cora Diamond, "Eating Meat and Eating People," *Philosophy* 53 (1978): 465–79.

24. Kant would agree.

25. Tom Regan, *The Case for Animal Rights* (Berkeley: University of California Press, 2004).

26. Richard Swinburne, *Providence and the Problem of Evil* (Oxford: Oxford University Press, 1998).

27. Regan, *The Case for Animal Rights*, 248.

28. Ibid., 243.

29. Ibid., 239.

30. Peter Singer, *Animal Liberation* (New York: HarperCollins, 2009).

31. Swinburne, *Providence and the Problem of Evil*, 171–75 and 189–92.

32. Regan, *The Case for Animal Rights,* 235.

33. Robert Nozick, *Anarchy, State, and Utopia* (New York: Basic Books, 1974).

34. Ibid., 45.

35. http://www.secondlivestock.com/public/vfr.php.

36. One might object that we are making this judgment from our standpoint—as creatures that value freedom (perhaps as one of our highest values)—but that things might look different from the standpoint of chickens themselves. But there is evidence that, from the standpoint of a chicken, they do prefer self-direction over the alternative: they are unhappy in cages, when bound, or when held. This seems to indicate that even from their standpoint, they value self-direction.

37. American Dietetic Association, "Position of the American Dietetic Association and Dietitians of Canada: Vegetarian Diets," *Journal of the American Dietetic Association* 103 (2003): 748.

38. For an interesting related discussion, see Kathryn Paxton George, "Should Feminists be Vegetarians?" *Signs: Journal of Women in Culture and Society* 19 (1994): 405–34.

39 Regan, *The Case for Animal Rights,* 308.

40. Ibid., 236–37.

41. Currently, this problem is difficult given the fact that if farmers don't use animal manure to fertilize, then they must use petroleum-based fertilizer that (a) strips the soil of its nutrients and (b) relies on oil production. One possibility is to use animal fertilizers without using the animals for food (or at least without killing them for food) while searching for alternative sustainable means of fertilizing crops.

Bibliography

A Big Love Story. Directed by Ryan Sage. 2012. Chatsworth, CA: Osiris Entertainment. 2013. DVD.

Adams, Carol. *The Sexual Politics of Meat: A Feminist Vegetarian Critical Theory*. New York: Continuum, 1990.

Adamson, Joni. "Medicine Food: Critical Environmental Justice Studies, Native North American Literature, and the Movement of Food Sovereignty." *Environmental Justice* 4 (2011): 213–19.

Adler-Hellman, Judith. *The World of Mexican Migrants: The Rock and the Hard Place*. New York: New Press, 2008.

Agarwal, Bina. "Food Crises and Gender Inequality." DESA Working Paper 107 (2011). Accessed March 8, 2015. http://www.un.org/esa/desa/papers/2011/wp107_2011.pdf.

Agarwal, Bina. *A Field of One's Own: Gender and Land Rights in South Asia*. Cambridge: Cambridge University Press, 1994.

Alaimo, Katherine, Christine M. Olson, and Edward A. Frongillo, Jr. "Family Food Insufficiency, but Not Low Family Income, is Positively Associated with Dysthymia and Suicide Symptoms in Adolescents." *Journal of Nutrition* 132 (2002): 719–25.

Alaimo, Katherine, Christine M. Olson, and Edward A. Frongillo, Jr. "Food Insufficiency and American School-Aged Children's Cognitive, Academic, and Psychosocial Development." *Pediatrics* 108 (2001): 44–53.

Alkon, Alison Hope and Julian Agyeman, editors. *Cultivating Food Justice: Race, Class, and Sustainability*. Cambridge, MA: MIT Press, 2011.

Allen, Arthur. "U.S. Touts Fruit and Vegetables While Subsidizing Animals that Become Meat." *Washington Post Health and Science*, October 3, 2011. http://www.washingtonpost.com/national/health-science/us-touts-fruit-and-vegetables-while-subsidizing-animals-that-become-meat/2011/08/22/gIQATFG5IL_story.html.

Allen, Patricia, and Carolyn Sachs. "Women and Food Chains: The Gendered Politics of Food." In *Taking Food Public: Redefining Foodways in a Changing World*, edited by Psyche Williams Forson and Carol Counihan. New York: Routledge, 2013.

Allen, Patricia, and Carolyn Sachs. "Women and Food Chains: The Gendered Politics of Food." *International Journal of Sociology of Food and Agriculture* 15, no. 1 (2007): 1–23.

Altidor, Paul. *Impacts of Trade Liberalization Policies on Rice Production in Haiti*. MIT Master's Thesis, 2004.

American Cancer Society, "What are the Risk Factors for Prostate Cancer?" last modified February 25, 2014, http://www.cancer.org/cancer/prostatecancer/overviewguide/prostate-cancer-overview-what-causes.

American Dietetic Association. "Position of the American Dietetic Association and Dietitians of Canada: Vegetarian Diets." *Journal of the American Dietetic Association* 103 (2003): 748–65.

American Meat Institute. "The United States Meat Industry at a Glance." Accessed June 15, 2014. http://www.meatami.com/ht/d/sp/i/47465/pid/47465.

American Time Use Survey, 2013. Table A-1. http://www.bls.gov/tus/.

Anderson, Elizabeth. "What is the Point of Equality?" *Ethics* 109 (1999): 287–337.

Angermeier, Paul L., and James R. Karr. "Biological Integrity versus Biological Diversity as Policy Directives." *BioScience* 44 (1994): 690–97.

Aquinas, Thomas. *On Law, Morality, and Politics*. Edited by William P. Baumgarth and Richard J. Regan. Indianapolis: Hackett, 1988.

Aristotle. *Nicomachean Ethics*. Translated by Terence Irwin. Indianapolis, Indiana: Hackett Publishing Company, 1985.

Aureli, Alice, and Claudine Brelet. "Water and Ethics, Women and Water: An Ethical Issue." United Nations Educational, Scientific and Cultural Organization, 2004, prepared by Water Aid, the Water Supply and Sanitation Collaboration Council, and Domestos. Accessed March 8, 2015. http://unesdoc.unesco.org/images/0013/001363/136357e.pdf.

Ball, Kylie, Anna Timperio, and David Crawford. "Neighborhood Socioeconomic Inequalities in Food Access and Affordability." *Health & Place* 15 (2009): 578–85.

Barrett, Christopher B., and Dan Maxwell. *Food Aid After Fifty Years: Recasting its Role*. New York: Routledge, 2007.

Barry, Brian. "Sustainability and Intergenerational Justice." In *Fairness and Futurity: Essays on Environmental Sustainability and Social Justice*, edited by Andrew Dobson, 93–117. Oxford: Oxford University Press, 1999. Originally published in *Theoria* 45 (June 1997): 43–65.

Bartky, Sandra Lee. *Femininity and Domination: Studies in the Phenomenology of Oppression*. New York: Routledge, 1990.

Bauer, Katherine W., Rachel Widome, John H. Himes, Mary Smyth, Bonnie Holy Rock, et al. "High Food Insecurity and Its Correlates Among Families Living on a Rural American Indian Reservation." *American Journal of Public Health* 102, no.7 (2012): 1346–52.

Baxter, Janeen, Belinda Hewitt, and Michele Haynes. "Life Course Transitions and Housework: Marriage, Parenthood, and Time on Housework." *Journal of Marriage and Family* 70 (2008): 259–72.

Beaulac, Julie, Elizabeth Kristjansson, and Steven Cummins. "A Systematic Review of Food Deserts, 1966–2007." *Preventing Chronic Disease* 6 (2009): A105.

Becker, Anne E. "Television, Disordered Eating, and Young Women in Fiji: Negotiating Body Image and Identity during Rapid Social Change." *Culture, Medicine and Psychiatry* 26 (2004): 533–59.

Becker, Anne E., Rebecca A. Burwell, Stephen Gilman, David B. Herzog, and Paul Hamburg. "Eating Behaviours and Attitudes following Prolonged Exposure to Television among Ethnic Fijian Adolescent Girls." *British Journal of Psychiatry* 180 (2002): 509–14.

Becker, Anne E., Rebecca A. Burwell, Kesaia Navara, and Steven E. Gilman. "Binge Eating and Binge Eating Disorder in a Small-Scale, Indigenous Society: The View from Fiji." *International Journal of Eating Disorders* 34 (2003): 423–31.

Becker, Lawrence C. "Against the Supposed Difference Between Historical and End-State Theories." *Philosophical Studies* 41 (1982): 267–72.

Bell, Richard A., Charles R. Berger, Diana Cassady, and Marilyn Townsend. "Portrayals of Food Practices and Exercise Behavior in Popular American Films." *Journal of Nutrition Education and Behavior* 37 (2005): 27–32.

Bello, Walden F. *The Food Wars*. London: Verso, 2009.

Beoku-Betts, Josephine A. "We Got Our Way of Cooking Things: Women, Food, and Preservation of Cultural Identity among the Gullah." *Gender & Society* 9 (1995): 535–55.

Berliner, Uri. "Haves and Have-Nots: Income Inequality in America." 2007. Accessed April 10, 2013. http://www.npr.org/templates/story/story.php?storyId=7180618.

Besky, Sarah. *The Darjeeling Distinction: Labor and Justice on Fair-Trade Tea Plantations in India.* Berkeley, CA: University of California Press, 2013.

Bittman, Mark. "Fast Food, Low Pay." *The New York Times*. July 26, 2013. A19.

Black, Christina, Georgia Ntani, Ross Kenny, Tannaze Tinati, Megan Jarman, Wendy Lawrence, Mary Barker, Hazel Inskip, Cyrus Cooper, Graham Moon, and Janis Baird. "Variety and Quality of Healthy Foods Differ According to Neighbourhood Deprivation." *Health & Place* 18 (2012): 1292–99.

Blake, Michael, and Patrick Taylor Smith. "International Distributive Justice." *Stanford Encyclopedia of Philosophy*, 2013. http://plato.stanford.edu/entries/international-justice/.

Block, Jason P., Richard A. Scribner, and Karen B. DeSalvo. "Fast Food, Race/Ethnicity, and Income." *American Journal of Preventive Medicine* 27, no.3 (2004): 211–17.

Bordo, Susan. *Unbearable Weight: Feminism, Western Culture, and the Body*. Tenth Anniversary Edition. Berkeley and Los Angeles: University of California Press, 2004.

Bordo, Susan. *Unbearable Weight: Feminism, Western Culture and the Body*. Berkeley: University of California Press, 1993.

Bower, Kelly M., Roland J. Thorpe Jr., Charles Rohde, and Darrell J. Gaskin. "The Intersection of Neighborhood Racial Segregation, Poverty, and Urbanicity and its Impact on Food Store Availability in the United States." *Preventative Medicine* 58 (2014): 33–9.

Braidotti, Rosi. *Nomadic Subjects*. New York: Columbia University Press, 1994.

Brodribb, Somer. *Nothing Mat(t)ers: A Feminist Critique of Postmodernism*. North Melbourne: Spinifex Press, 1992.

Budzynska, Katarzyna, Patricia West, Ruth T. Savoy-Moore, Darlene Lindsey, Michael Winter, and P. K. Newby. "A Food Desert in Detroit: Associations with Food Shopping and Eating Behaviors, Dietary Intakes and Obesity." *Public Health Nutrition* 16 no. 2 (2013): 2114–23.

Bullard, Robert D. *Dumping in Dixie: Race, Class, and Environmental Quality*. Boulder, CO: Westview Press, 1990.

Butler, Judith. "What is Critique? An Essay on Foucault's Virtue." In *The Political: Readings in Continental Philosophy*, edited by David Ingram. Oxford: Blackwell, 2002.

Butterly, John R., and Jack Shepherd. *Hunger: The Biology and Politics of Starvation*. Hanover, NH: Dartmouth College Press, 2010.

Buzby, Jean C., Honan Farah Wells, and Jeffrey Hyman. "The Estimated Amount, Value, and Calories of Postharvest Food Losses at the Retail and Consumer Levels in the United States." USDA Report, 2014.

Callicott, J. Baird. "Animal Liberation: A Triangular Affair." *Environmental Ethics* 2 (1980): 311–28.

Callicott, J. Baird. "The Value of Ecosystem Health." In J. Baird Callicott, *Beyond the Land Ethic: More Essays in Environmental Philosophy*, 347–64. Albany: State University of New York Press, 1999. Originally published in *Environmental Values* 5 (1995): 345–61.

Carrigan, Ana. "Chiapas, The First Postmodern Revolution." In *Our Word is Our Weapon: Selected Writings, Subcomandante Insurgente Marcos,* edited by Juana Ponce de Leon. New York: Seven Stories Press, 2001.

Chavez, Manuel. "Asymmetry of Resources, Access to Information, and Transparency as Structural Development Challenges in Rural Areas." In *NAFTA and the Campesinos: The Impact of NAFTA on Small-Scale Agricultural Producers in Mexico and the Prospects for Change*, edited by Juan Rivera, Scott Whiteford, and Manuel Chavez. Scranton: University of Scranton Press, 2008.

Chef. Directed by Jon Favreau. 2014. Hollywood: Open Road Films. DVD.

Chernin, Kim. *The Obsession: Reflections on the Tyranny of Slenderness*. New York: Harper and Row, 1981.

Chisuwa, Naomi, and Jennifer A. O'Dea. "Body Image and Eating Disorders amongst Japanese Adolescents: A Review of the Literature." *Appetite* 54 (2010): 5–15.

Chocolat. Directed by Lasse Hallstrom. 2000. Santa Monica: Miramax Lionsgate. 2011. DVD.

Chu, Kathy. "Extreme Dieting Spreads to Asia." *USA Today*. March 29, 2010.

Claeys, Priscilla. "From Food Sovereignty to Peasants' Rights: An Overview of La Via Campesina's Rights-Based Claims Over the Last 20 Years." *Food Sovereignty: A critical dialogue*. International Conference, Yale University, 2013.

Coleman-Jensen, Alisha, and Mark Nord. "Disability is an Important Risk Factor for Food Insecurity." USDA Economic Research Service, 2013. Accessed March 25, 2015. http://www.ers.usda.gov/amber-waves/2013-may/disability-is-an-important-risk-factor-for-food-insecurity.aspx#.VRLO946jOSo.

Coleman-Jensen, Alisha, Christian Gregory, and Anita Singh. *Household Food Security in the United States in 2013: Statistical Supplement*. United States Department of Agriculture. Economic Research Service. AP-066. September 2014. Accessed March 11, 2015. http://www.ers.usda.gov/media/1565530/ap066.pdf.

Coleman-Jensen, Alisha, Christian Gregory, and Anita Singh, "Household Food Security in the United States in 2013." United States Department of Agriculture Economic Report, September 2014.

Coleman-Jensen, Alisha, William McFall, and Mark Nord. "Food Insecurity in Households with Children: Prevalence, Severity, and Household Characteristics, 2010–11." USDA Report Summary, 2013.

Collier, Paul. "The Politics of Hunger: How Illusion and Greed Fan the Food Crisis." *Foreign Affairs* 87 (2008): 67–79.

Community Development Advocates of Detroit. "What We Do." Cdad-online.org (2015).

Cook, John T., Deborah A. Frank, Carol Berkowitz, Maureen M. Black, Patrick H. Casey, Diana B. Cutts, Alan F. Meyers, Nieves Zaldivar, Anne Skalicky, Suzette Levenson, Tim Heeren, and Mark Nord. "Food Insecurity is Associated with Adverse Health Outcomes Among Human Infants and Toddlers." *The Journal of Nutrition* 134 (2004): 1432–38.

Council of Canadian Academics. "Aboriginal Food Security in Northern Canada: An Assessment of the State of Knowledge." Accessed May 4, 2015. http://www.scienceadvice.ca/en/assessments/completed/food-security.aspx.

Counihan, Carole. *Around the Tuscan Table: Food, Family, and Gender in Twentieth-Century Florence*. New York: Routledge, 2004.

Counihan, Carole M. *The Anthropology of Food and Body: Gender, Meaning, and Power*. New York: Routledge, 1999.

Courtenay, Will. "Behavioral Factors Associated with Disease, Injury, and Death among Men: Evidence and Implications for Prevention." *The Journal of Men's Studies* 9 (2000): 81–111.

Daly, Herman. "Sustainable Economic Development: Definitions, Principles, Policies." In *The Essential Agrarian Reader: The Future of Culture, Community, and the Land*, edited by Norman Wirzba, 62–79. Lexington: University Press of Kentucky, 2003.

Daly, Herman. "On Wilfred Beckerman's Critique of Sustainable Development." *Environmental Values* 4 (1995): 49–55.

Davis, Mike. *Planet of Slums*. Reprint edition. New York: Verso, 2007.

Declaration of Nyéléni. Adopted at the Forum for Food Sovereignty, Sélingué, Mali, February 27, 2007. Accessed June 14, 2014. http://www.nyeleni.org/spip.php?article290.

Delivering the Goods. Directed by Matthew Bonifacio. 2011. Beverley Hills, CA: GoDigital. 2012.

Desmarais, Annette Aurélie. "The Power of Peasants: Reflections on the Meanings of La Vía Campesina." *Journal of Rural Studies* 24 (2008): 138–49.

Desmarais, Annette Aurélie. *La Vía Campesina: Globalization and the Power of Peasants*. Fernwood Publishing, 2007.

Desmarais, Annette Aurélie. "The Via Campesina: Peasant Women at the Frontiers of Food Sovereignty." *Canadian Woman Studies* 23 (2003): 140–45.

Detroit Food Justice Task Force. "Food Justice." http://www.detroitfoodjustice.org/.

Detroit People's Platform. "Community Land Trust." Detroitpeoplesplatform.org (2015).

Detroit People's Platform. "Land." Detroitpeoplesplatform.org (2015).

Devasahayam, Theresa W. "Power and Pleasure around the Stove: The Construction of Gendered Identity in Middle-Class South Indian Hindu Households in Urban Malaysia." In *Women's Studies International Forum*, 28:1–20. Elsevier, 2005.

Diamond, Cora. "Eating Meat and Eating People." *Philosophy* 53 (1978): 465–79.

Disfigured. Directed by Glenn Gers. 2008. Canoga Park, CA: Cinema Libre. DVD.

Dohnt, Hayley K., and Marka Tiggemann. "Body Image Concerns in Young Girls: The Role of Peers and Media Prior to Adolescence." *Journal of Youth and Adolescence* 35 (2006): 141–51.

Domestos, Water Aid, and WSSCC. "Why We Can't Wait: A Report on Sanitation and Hygiene for Women and Girls." Accessed March 8, 2015. http://www.zaragoza.es/contenidos/medioambiente/onu/1325-eng_We_cant_wait_sanitation_and_hygiene_for_women_and%20girls.pdf.

Dowie, Mark. *Conservation Refugees: The Hundred-Year Conflict between Global Conservation and Native Peoples*. Cambridge: MIT Press, 2009.

Dube, Laurette, Jordan L. LeBel, and Ji Lu. "Affect Asymmetry and Comfort Food Consumption." *Physiology & Behavior* 86 (2005): 559–67.

Dutko, Paula, Michele Ver Ploeg, and Tracey Farrigan. "Characteristics and Influential Factors of Food Deserts." USDA Economic Research Report 140 (2012).

Eat, Drink, Man, Woman. Directed by Ang Lee. 1994. Beverly Hills: MGM World Films. 2002.

Editors. "Eating With Our Eyes Closed." *The New York Times*. April 10, 2012. A20.

Editors. "From The Editors." *Race/Ethnicity* 5 (2011): vii–xvii.

Edkins, Jenny. *Whose Hunger? Concepts of Famine, Practices of Aid*. Minneapolis: University of Minnesota Press, 2000.

Ellingwood, Ken. *Hard Line: Life and Death on the U.S.–Mexico Border*. New York: Random House, 2005.

Elliot, Robert. *Faking Nature: The Ethics of Environmental Restoration*. London: Routledge, 1997.

Elmhirst, R. "Introducing New Feminist Political Ecologies." *Geoforum* 42 (2011): 129–32.

El Serafy, Salah. "In Defence of Weak Sustainability: A Response to Beckerman." *Environmental Values* 5 (1996): 75–81.

European Commission. "Food Waste." Last updated March 9, 2015. http://ec.europa.eu/food/safety/food_waste/index_en.htm.

FAO, IFAD, and WFP. *The State of Food Insecurity in the World 2014: Strengthening the Enabling Environment for Food Security and Nutrition*. Rome: FAO, 2014.

FAO, WFP, and IFAD. *The State of Food Insecurity in the World 2012: Economic Growth is Necessary but not Sufficient to Accelerate Reduction of Hunger and Malnutrition*. Rome: FAO, 2012.

"FAO Summit Ignored Women Farmers." *Appropriate Technology* 35 (2008): 6.

Fat Girl. Directed by Catherine Breillat. 2001. New York: Criterion. 2004. DVD.

Feldman, Shelley. "Rethinking Development, Sustainability, and Gender Relations." *Cornell Journal of Law & Public Policy* 22 (2012): 649.

Feldman, Shelley, and Stephen Biggs. "International Shifts in Agricultural Debates and Practice: An Historical View of Analyses of Global Agriculture." In *Integrating Agriculture, Conservation and Ecotourism: Societal Influences*, edited by W. Bruce Campbell and Silvia López Ortíz, 107–61. Issues in Agroecology—Present Status and Future Prospectus 2. Springer Netherlands, 2012.

Flora, C. "Book Review: Schanbacher, William D. The Politics of Food: The Global Conflict between Food Security and Food Sovereignty." *Journal of Agricultural and Environmental Ethics* 24 (2010): 545–47.

Food and Agriculture Organization of the United Nations. "FAO in Emergencies Guidance Note: Striving for Gender Equality in Emergencies." Rome: Food and Agricultural Organization of the United Nations, 2013. http://www.fao.org/fileadmin/user_upload/emergencies/docs/Guidance%20Note%20Gender.pdf.

Food and Agriculture Organization of the United Nations. "Food Security." Rome: Food and Agricultural Organization of the United Nations Policy Brief, 2006.

Food and Agriculture Organization of the United Nations. "Food Security: Concepts and Measurement." Accessed July 14, 2014. http://www.fao.org/docrep/005/y4671e/y4671e06.htm.

Food and Agricultural Organization of the United Nations. "Food Security Statistics." Accessed May 7, 2015. http://www.fao.org/economic/ess/ess-fs/en/

Food and Agriculture Organization of the United Nations. "Gender," 2015. http://www.fao.org/gender/gender-home/en/?no_cache=1.

Food and Agriculture Organization of the United Nations. "Gender Equality and Food Security." Rome: Food and Agricultural Organization of the United Nations, 2013. Accessed June 19, 2014. http://www.fao.org/wairdocs/ar259e/ar259e.pdf

Food and Agriculture Organization of the United Nations. "Global Agriculture Towards 2050." Rome: Food and Agricultural Organization of the United Nations, 2009.

Food and Agriculture Organization of the United Nations. "10 Hunger Facts for 2014." Rome: Food and Agricultural Organization of the United Nations, 2013. Accessed 19 June, 2014. https://www.wfp.org/stories/10-hunger-facts-2014.

Food and Agriculture Organization of the United Nations. "Passport to Mainstreaming Gender in Water Programmes: Key Questions for Interventions in the Agricultural Sector." Food and Agricultural Organization of the United Nations, 2012. Accessed March 8, 2015. http://www.fao.org/docrep/017/i3173e/i3173e.pdf.

Food and Agriculture Organization of the United Nations. "Trade Reforms and Food Security: Conceptualizing the Linkages." Rome: Food and Agricultural Organization of the United Nations, 2003.

Food and Agriculture Organization of the United Nations. *World Food Summit: Food For All.* Rome: Food and Agricultural Organization of the United Nations, 1996.

Food Desert Website. "Maps Page." Accessed November 20, 2014. www.fooddeserts.org/images/kontentMaps.htm.

Food, Inc. Directed by Robert Kenner. 2008. Participant Media.

Food Trust. "What We Do." Accessed November 16, 2014. http://thefoodtrust.org/what-we-do/supermarkets.

Forde, Steven. "The Charitable John Locke." *The Review of Politics* 71 (2009): 428–58.

Foucault, Michel. "Body/Power and Truth and Power." In *Michel Foucault: Power/Knowledge*, edited by C. Gordon. U.K. Harvester, 1980.

Foucault, Michel. *Discipline and Punishment*. Translated by Alan Sheridan. New York: Pantheon, 1977.

Foucault, Michel. "The Ethics of the Concern for Self as a Practice of Freedom." In *The Essential Works of Foucault 1954–1984: Vol. 1: Ethics, Subjectivity and Truth,* edited by Paul Rabinow. New York: New Press, 1997.

Foucault, Michel. *The History of Sexuality, Vol. 1: An Introduction*. Translated by Robert Hurley. New York: Vintage, 1980.

Foucault, Michel. "Qu'est-ce que la critique?" *Bulletin de la Societe Francaise de Philosophie* 84 (1990): 35–63.

Fraser, Arabella. "Harnessing Agriculture for Development." *Oxfam Policy and Practice: Agriculture, Food and Land* 9 (2009): 56–130.

Fraser, Nancy. *Justice Interruptus: Critical Reflections on the "Postsocialist" Condition.* New York: Routledge, 1997

Fraser, Nancy, and Axel Honneth. *Redistribution or Recognition: A Political Philosophical Exchange*. New York: Verso, 2003.

Freeman, Samuel. *Rawls*. New York: Routledge, 2007.

Friedmann, Harriet. "From Colonialism to Green Capitalism: Social Movements and Emergence of Food Regimes." *Research in Rural Sociology and Development* 11 (2005): 227–64.

Friedmann, Harriet. "The Political Economy of Food: A Global Crisis." *New Left Review* (1993): 29–57.

Friedmann, Harriet. "The Political Economy of Food: The Rise and Fall of the Postwar International Food Order. *American Journal of Sociology* (1982) S248-S286.

Frontline. "Antibiotic Debate Overview." Accessed July 28, 2013. ww.pbs.org/wgbh/pages/frontline/shows/meat/safe/overview.html.

Frye, Marilyn. *The Politics of Reality: Essays in Feminist Thought.* California: The Crossing Press, 1983.

Gaesser, Glenn A. *Big Fat Lies*. New York: Fawcett Columbine, 1996.

Gallagher, Mari. "Food Desert and Food Balance Community Fact Sheet." Mari Gallagher Research and Consulting Group (2010). Accessed June 1, 2013. http://www.fooddesert.net/wp-content/themes/cleanr/images/FoodDesertFactSheet-revised.pdf.

Gallagher, Mari. "Examining the Impact of Food Deserts on Public Health in Chicago." Mari Gallagher Research and Consulting Group (2006).

Gallagher, Mari. "Examining the Impact of Food Deserts on Public Health in Detroit." Mari Gallagher Research Group (2007).

"Gender Equality and Food Security: Women's Empowerment as a Tool against Hunger." Asian Development Bank, 2013. Accessed March 8, 2015. http://www.

adb.org/publications/gender-equality-and-food-security-womens-empowerment-tool-against-hunger.

Gomberg-Munoz, Ruth. *Labor and Legality: An Ethnography of a Mexican Immigrant Network*. New York: Oxford University Press, 2011.

Gomez Cruz, Manuel Angel, and Rita Schwentesius Rindermann. "NAFTA's Impact on Mexican Agriculture: An Overview." In *NAFTA and the Campesinos: The Impact of NAFTA on Small-Scale Agricultural Producers in Mexico and the Prospects for Change*, edited by Juan Rivera, Scott Whiteford, and Manuel Chavez. Scranton: University of Scranton Press, 2008.

Gould, Carol. *Interactive Democracy: The Social Roots of Global Justice*. Cambridge, UK: Cambridge University Press, 2014.

Gould, Carol C. *Rethinking Democracy: Freedom and Social Cooperation in Politics, Economy, and Society*. Cambridge: Cambridge University Press, 1988.

Gray, Kevin. "Going Vegan in the NFL." *Men's Journal*, December 2012. http://www.mensjournal.com/magazine/going-vegan-in-the-nfl-20130123.

Great Vegan Athletes. Accessed March 23, 2014. http://www.greatveganathletes.com/vegan_athlete_patrik-baboumian-vegan-strongman and http://www.greatveganathletes.com/vegan_athlete_billy-simmonds-vegan-bodybuilder.

Greene Sterline, Terry. *Illegal: Life and Death in Arizona's Immigration War Zone*. Guilford: Lyons Press, 2010.

Grijalva, James M. "Self-Determining Environmental Justice for Native America." *Environmental Justice* 4 (2011): 187–92.

Grown, Caren. "Missing Women: Gender and the Extreme Poverty Debate." USAID, 2014. http://usaidlearninglab.org/sites/default/files/resource/files/Gender%20&%20Extreme%20Poverty_Missing%20Women.pdf.

Guayaki. "The Guayaki Story." Accessed November 17, 2014. http://guayaki.com/about/134/The-Guayak%26iacute;-Story.html.

Guha, Ramachandra, and Juan Martinez-Alier. *Varieties of Environmentalism: Essays North and South*. London: Earthscan Publications Ltd., 1997.

Guthman, Julie. "'If Only They Knew': The Unbearable Whiteness of Alternative Food." In *Cultivating Food Justice: Race, Class, and Sustainability*, edited by Alison Hope Alkon and Julian Agyeman, 263–81. Cambridge, MA: MIT Press, 2011.

Hanlon, Joseph, Armando Barrientos, and David Hulme. *Just Give Money to the Poor: The Development Revolution from the Global South*. Sterling, VA: Kumarian Press, 2010.

Hannah and Her Sisters. Directed by Woody Allen. 1986. Los Angeles: 20th Century Fox. 2001.

Hartsock, Nancy. *The Feminist Standpoint Revisited & Other Essays*. Boulder: Westview Press, 1998.

Hassoun, Nicole. *Globalization and Global Justice: Shrinking Distance, Expanding Obligations*. Cambridge: University Press Cambridge, 2012.

Ha-Won, Jung. "South Korea's Plastic Surgery Fad Goes Extreme." *Jakarta Globe*. May 27, 2013. Accessed April 22, 2015. http://www.thejakartaglobe.com/features/south-koreas-plastic-surgery-fad-goes extreme/?doing_wp_cron=1373226815.186551094055175781 2500.

Hemingway, Ernest. *A Moveable Feast.* New York: Scribner, 2009.

Hendrickson, Deja, Cherry Smith, and Nicole Eikenberry. "Fruit and Vegetable Access in Four Low-Income Food Deserts Communities in Minnesota." *Agriculture and Human Values* 23 (2006): 371–83.

Henning, Brian G. "Standing in Livestock's 'Long Shadow': The Ethics of Eating Meat on a Small Planet." *Ethics and the Environment* 16 (2011): 63–93.

Heyes, Cressida. *Self-Transformations: Foucault, Ethics, and Normalized Bodies.* Oxford: Oxford University Press, 2007.

Hickson, Meredith, Stephanie Ettinger de Cuba, Ingrid Weiss, Gemma Donofrio, and John Cook. "Feeding Our Human Capital: Food Insecurity and Tomorrow's Workforce." Children's Health Watch Research Brief (2013). Accessed July 15, 2014. http://www.childrenshealthwatch.org/wp-content/uploads/FeedingHumanCapital_report.pdf.

Hickson, Meredith, Stephanie Ettinger de Cuba, Ingrid Weiss, Gemma Donofrio, and John Cook. "Too Hungry to Learn: Food Insecurity and School Readiness." Children's Health Watch Research Brief (2013). Accessed July 15, 2014. http://www.childrenshealthwatch.org/wp-content/uploads/toohungrytolearn_report.pdf.

Hochschild, Arlie, and Anne Machung. *The Second Shift: Working Parents and the Revolution at Home*. New York: Viking Press, 1989.

The Holiday. Directed by Nancy Meyers. 2006. Culver City, CA: Sony Pictures Home Entertainment, 2007. DVD.

Holland, Alan. "Ecological Integrity and the Darwinian Paradigm." In *Ecological Integrity: Integrating Environment, Conservation, and Health*, edited by David Pimental, Laura Westra, and Reed F. Noss, 45–59. Washington, D.C.: Island Press, 2000.

Holland, Alan. "Substitutability: Or, Why Strong Sustainability is Weak and Absurdly Strong Sustainability is Not Absurd." In *Valuing Nature? Ethics, Economics and the Environment*, edited by John Foster, 119-34. London: Routledge, 1997.

Holland, Alan. "Sustainability: Should We Start From Here?" In *Fairness and Futurity: Essays on Environmental Sustainability and Social Justice*, edited by Andrew Dobson, 46–68. Oxford: Oxford University Press, 1999.

Holmes, Seth. *Fresh Fruit, Broken Bodies: Migrant Farmworkers in the United States.* Los Angeles: University of California Press, 2013.

Hoover, E. "Cultural and Health Implications of Fish Advisories in a Native American Community." *Ecological Processes* 2 (2013): 4.

Hursthouse, Rosalind. *On Virtue Ethics*. Oxford: Oxford University Press, 1999.

Imhoff, Daniel. "Overhauling the Farm Bill: The Real Beneficiaries of Subsidies." *The Atlantic*. 2012. www.theatlantic.com/health/archive/2012/03/overhauling-the-farm-bill-the-real-beneficiaries-of-subsidies/254422/.

In Her Skin. Directed by Simone North. 2009. Orland Park, IL: MPI Home Video. 2011. DVD.

Institute of Medicine. "Supplemental Nutrition Assistance Program: Examining the Evidence to Define Benefit Adequacy." January 17, 2013. Accessed March 10, 2015. http://www.iom.edu/snapadequacy.

Jacobs, Michael. "Sustainable Development, Capital Substitution and Economic Humility: A Response to Beckerman." *Environmental Values* 4 (1995): 57–68.

Jacobson, Rebecca. "The Bitter Taste of Genetics." *PBS NewsHour*. December 23, 2010. Accessed April 22, 2015. http://www.pbs.org/newshour/updates/science-july-dec10-geneticstaste_12-23/.

Jaggar, Alison M., "'Saving Amina': Global Justice for Women and Intercultural Dialogue." *Ethics & International Affairs* 19 (2005): 55–75.

Jaggar, Alison M. and Iris Marion Young, editors. *A Companion to Feminist Philosophy*. Malden, MA.: Blackwell Publishers Inc., 1998.

Jenkins, Alan. "Inequality, Race, and Remedy." *The American Prospect*. 2007. Accessed April 11, 2013. http://prospect.org/article/inequality-race-and-remedy.

Just Food. http://www.justfood.org/.

Jyoti, Diana F., Edward A. Frongillo, and Sonya J. Jones. "Food Insecurity Affects School Children's Academic Performance, Weight Gain, and Social Skills." *Journal of Nutrition* 135 (2005): 2831–39.

Karpyn, Allison, and Sarah Treuhaft. "The Grocery Gap: Finding Healthy Food in America." In *A Place at the Table*, edited by Peter Pringle, 45–58. New York: Public Affairs, 2013.

Karr, James R. "Ecological Integrity and Ecological Health Are Not the Same." In *Engineering Within Ecological Constraints*, edited by Peter C. Schulze, 97–109. Washington, D.C.: National Academy Press, 1996.

Karr, James R. "Health, Integrity, and Biological Assessment: The Importance of Measuring Whole Things." In *Ecological Integrity: Integrating Environment, Conservation, and Health*, edited by David Pimental, Laura Westra, and Reed F. Noss, 209–26. Washington, D.C.: Island Press, 2000.

Kasperson, R. E. "Six Propositions for Public Participation and Their Relevance for Risk Communication." *Risk Analysis* 6 (1986): 275–81

Khadse, Ashlesha and Niloshree Bhattacharya. "India: A Conversation with Farmers of the KRRS." *La Via Campesina's Open Book: Celebrating 20 Years of Struggle and Hope*. Viacampesina.org, 2013.

The Kids Are Alright. Directed by Lisa Cholodenko. 2010. Universal City, CA: Focus Features. DVD.

Kilcullen, John. "The Origin of Property: Ockham, Grotius, Pufendorf, and Some Others." In *A Translation of William of Ockham's Work of Ninety Days*, translated by J. Kilcullen and J. Scott. Lewiston, NY: Edwin Mellon Press, 2001.

Kimmel, Michael. "Masculinity as Homophobia: Fear, Shame, and Silence in the Construction of Gender Identity." In *Theorizing Masculinities*, edited by Harry Brod and Michael Kaufman, 119–42, Research on Men and Masculinities Series. Thousand Oaks, CA: Sage Publications, Inc., 1994.

Koeth, Robert A., et al. "Intestinal Microbiota Metabolism of L-Carnitine, a Nutrient in Red Meat, Promotes Atherosclerosis." *Nature Medicine* 19 (2013): 576–85. Accessed July 13, 2014. doi: 10.1038/nm.3145.

Korsmeyer, Carolyn. *Making Sense of Taste: Food and Philosophy*. Ithaca: Cornell University Press, 1999.

Kraut, Richard. *What Is Good and Why: The Ethics of Well-Being*. Cambridge, MA: Harvard University Press, 2007.

LaFollette, Hugh. *World Hunger*. Malden, MA: Blackwell, 2003.

LaFollette, Hugh, and William Aiken, editors. *World Hunger and Morality*. Englewood Cliffs: Prentice Hall, 1996.

Lamont, Julian, and Christi Favor. "Distributive Justice." *Stanford Encyclopedia of Philosophy*, 2013. http://plato.stanford.edu/entries/justice-distributive/.

Larsen, Kristian, and Jason Gilliland. "Mapping the Evolution of 'Food Deserts' in a Canadian City: Supermarket Accessibility in London, Ontario, 1961–2005." *International Journal of Health Geographics* 7 (2008): 16.

Last Holiday. Directed by Wayne Wang. 2006. Los Angeles: Warner Brothers. DVD.

La Via Campesina. "Declaration of Nyéléni." *La Via Campesina International Peasant's Movement*, February 27, 2007. http://viacampesina.org/en/index.php/main-issues-mainmenu-27/food-sovereignty-and-trade-mainmenu-38/262-declaration-of-nyi.

La Via Campesina. "The International Peasant's Voice." *La Via Campesina International Peasant's Movement*, February 9, 2011. http://viacampesina.org/en/index.php/organisation-mainmenu-44.

La Via Campesina. "Our World is Not For Sale—Priority to People's Food Sovereignty, WTO Out of Agriculture." viacampesina.org, 2001.

La Via Campesina. "Tlaxcala Declaration of the Via Campesina." viacampesina.org, 1996.

La Via Campesina. *Tlaxcala Declaration of La Vía Campesina*. Tlaxcala, Mexico, 1996. http://viacampesina.org/en/index.php/our-conferences-mainmenu-28/2-tlaxcala-1996-mainmenu-48/425-ii-international-conference-of-the-via-campesina-tlaxcala-mexico-april-18-21.

Lear, Jonathan. *Radical Hope: Ethics in the Face of Cultural Devastation*. Cambridge, MA: Harvard University Press, 2006.

Lelkes, Orsolva, and Eszter Zólyomi. "Poverty and Social Exclusion of Migrants in the European Union." European Centre Policy Brief, March 2011.

Lelwica, Michelle, Emma Hoglund, and Jenna McNallie. "Spreading the Religion of Thinness from California to Calcutta: A Critical Feminist Postcolonial Analysis." *Journal of Feminist Studies in Religion* 25 (2009): 19–41.

Leopold, Aldo. *A Sand County Almanac and Sketches Here and There*. New York: Oxford University Press, Inc., 1949.

"Ley de Seguridad Alimentaria y Nutricional." legislacion.asamblea.gob.ni, 2009.

Levant, R. F., R. Wu and J. Fischer, "Masculinity Ideology: A Comparison Between U.S. and Chinese Young Men and Women," *Journal of Gender, Culture and Health* 1(1996): 207–220.

Levesque, Maurice J., and David R. Vichesky. "Raising the Bar on the Body Beautiful: An Analysis of the Body Image Concerns of Homosexual Men." *Body Image* 3 (2006): 45–55.

Levinas, Emmanuel. *God, Death, and Time*. Translated by Bettina Bergo. Stanford, CA: Stanford University Press, 2000.

Levinas, Emmanuel. *Nine Talmudic Readings*. Translated by Annette Aronowicz. Bloomington, IN: Indiana University Press, 1990.

Levinas, Emmanuel. *Otherwise than Being or Beyond Essence*. Translated by Alphonso Lingis. Pittsburgh: Duquesne University Press, 1998.

Levinas, Emmanuel. "Secularization and Hunger." Translated by Bettina Bergo. *Graduate Faculty Philosophy Journal* 20/2–21/1 (1998): 3–12.

Levinas, Emmanuel. *Totality and Infinity*. Translated by Alphonso Lingis. Pittsburgh: Duquesne University Press, 1969.

Levitan, Robert D., and Caroline Davis. "Emotions and Eating Behaviour: Implications for the Current Obesity Epidemic." *University of Toronto Quarterly* 79 (2010): 783–99.

Lewin, Kurt. "Forces behind Food Habits and Methods of Change." In *The Problem of Changing Food Habits*. Bulletins of the National Research Council 108. Washington D.C.: National Research Council, National Academy of Science, 1943. http://www.nap.edu/openbook.php?record_id=9566&page=35.

Like Water for Chocolate. Directed by Alfonso Arau. 1992. Santa Monica, CA: Miramax Lionsgate. 2011. DVD.

Lintott, Sheila. "Sublime Hunger: A Consideration of Eating Disorders Beyond Beauty." *Hypatia* 18 (2003): 65–86.

Lo, Joann. "Racism, Gender Discrimination, and Food Chain Workers in the United States." In *Global Food Systems: The Issues and Solutions*, edited by William D. Schanbacher. ABC-CLIO, 2014.

Locke, John. "Draft of a Representation Containing a Scheme of Methods for the Employment of the Poor." In *John Locke: Political Writings*, edited by David Wootton. Indianapolis: Hackett, 1993.

Locke, John. *The First Treatise of Government*. In *Two Treatises of Government*, by John Locke, edited by Peter Laslett. Cambridge: Cambridge University Press, 1988.

Locke, John. *The Second Treatise of Government*. Edited by C. B. Macpherson. Indianapolis: Hackett, 1980.

Locke, John. *Two Treatises of Government*. Introduction and notes by Peter Laslett. New York: Cambridge University Press, 1963.

Locke, John. "Venditio." In *John Locke: Political Writings*, edited by David Wootton. Indianapolis: Hackett, 1993.

Love Food Hate Waste. "The Facts About Food Waste." Accessed April 20, 2015. http://england.lovefoodhatewaste.com/node/2472.

Lundberg, Shelly J., and Richard Startz. "Inequality and Race: Models and Policy." 1998. Accessed April 10, 2013. http://www.econ.washington.edu/user/startz/Working_Papers/Inequality.PDF.

Macht, Michael. "How Emotions Affect Eating: A Five-Way Model." *Appetite* 50 (2008): 1–11.

Macpherson, C. B. Introduction to *Second Treatise of Government*, by John Locke. Indianapolis: Hackett, 1980.

Macpherson, C. B. *The Political Theory of Possessive Individualism: Hobbes to Locke*. Oxford: Clarendon Press, 1962.

Makino, M., M. Hashizume, K. Tsuboi, M. Yasushi, and L. Dennerstein. "Comparative Study of Attitudes to Eating between Male and Female

Students in the People's Republic of China." *Eating Weight Disorder* 11 (2006): 111–17.

Martin, Adrian, Shawn McGuire, and Sian Sullivan. "Global Environmental Justice and Biodiversity Conservation." *The Geographical Journal* 179 (2013): 122–31.

Masson, Jeffrey Moussaieff. *The Face on Your Plate: The Truth about Food.* New York: W. W. Norton & Company, Inc., 2009.

May, Larry. *Sharing Responsibility.* Chicago: University of Chicago Press, 1992.

McClintock, Nathan. "From Industrial Garden to Food Desert: Demarcated Devaluation in the Flatlands of Oakland, California." In *Cultivating Food Justice: Race, Class, and Sustainability*, edited by Alison Hope Alkon and Julian Agyeman, 89–120. Cambridge: MIT Press, 2013.

McLaren, Margaret A. *Feminism, Foucault, and Embodied Subjectivity.* New York: State University of New York Press, 2002.

McLean, Bethany. "Food Deserts in America, Illustrated." 2011. Accessed April 10, 2013. http://www.huffingtonpost.com/2011/01/04/food-deserts-map_n_804110.html.

McMichael, P. D. "Global Development and the Corporate Food Regime." *Research in Rural Sociology and Development* 11 (2005): 265–63.

McMichael, P. D. "Tensions Between National and International Control of the World Food Order: Contours of a New Food Regime." *Sociological Perspectives* 35 (1992): 343–65.

McWilliams, James E. "The Myth of Sustainable Meat." *New York Times.* April 12, 2012.

McWilliams, James E. *Just Food: Where Locavores Get it Wrong and How We Can Truly Eat Responsibly.* New York: Little, Brown and Company, 2009.

Medical News Today. "Why Men Are More Prone To Heart Disease: New Research Led By University Of Leicester." September 1, 2008, http://www.medicalnewstoday.com/releases/119844.php.

Menser, Michael. "Transnational Participatory Democracy in Action: The Case of La Via Campesina." *Journal of Social Philosophy* 39 (2008): 20–41.

Menser, Michael. "Transnational Self-Determination, Food Sovereignty, and the State." In *Globalization and Food Sovereignty: Global and Local Change in the New Politics of Food,* edited by Peter Andree, Jeffrey Ayers, Michael Bosia, and Marie-Joesee Massicotte. Toronto: University of Toronto Press, 2010.

Mies, Maria, and Vandana Shiva. *Ecofeminism.* 2nd edition. Zed Books, 2014.

Miller, David. "Social Justice and Environmental Goods." In *Fairness and Futurity: Essays on Environmental Sustainability and Social Justice*, edited by Andrew Dobson, 151–72. Oxford: Oxford University Press, 1999.

Mills, Charles W. "'Ideal Theory' as Ideology." *Hypatia* 20 (2005): 165–83.

The Mistress of Spices. Directed by Paul Mayeda Berges. 2005. New York: Weinstein Company. 2007. DVD.

Mize, Ronald, and Alicia Swords. *Consuming Mexican Labor: From the Bracero Program to NAFTA.* Toronto: University of Toronto Press, 2011.

Mohai, Paul, David Pellow, and J. Timmons Roberts. "Environmental Justice." *Annual Review of Environment and Resources* 34 (2009): 405–30.

Moore, Harriett Bruce. "The Meaning of Food." *The American Journal of Clinical Nutrition* 5 (1957): 77–82.

Morland, K., S. Wing, A. Diez Roux, and C. Poole. "Neighborhood Characteristics Associated with the Location of Food Stores and Food Service Places." *American Journal of Preventative Medicine* 22 (2002): 23–29.

Morris, Michael, Valerie A. Kelly, Ron J. Kopicki, and Derek Byerlee. *Fertilizer Use in African Agriculture: Lessons Learned and Good Practice Guidelines*. Washington, D.C.: World Bank, 2007. Accessed March 10, 2015. http://documents.worldbank.org/curated/en/2007/01/7462470/fertilizer-use-african-agriculture-lessons-learned-good-practice-guidelines.

Morrison, Melanie A., Todd G. Morrison, and Cheryl-Lee Sager. "Does Body Satisfaction Differ between Gay Men and Lesbian Women and Heterosexual Men and Women? A Meta-Analytic Review." *Body Image* 1 (2004): 127–38.

Morrison, Todd G., and Marie Halton. "Buff, Tough, and Rough: Representations of Muscularity in Action Motion Pictures." *The Journal of Men's Studies* 17 (2009): 57–74.

Mostly Martha. Directed by Sandra Nettelbeck. 2001. Hollywood: Paramount Classics. 2003. DVD.

Mullany, Brita, Nicole Neault, Danielle Tsingine, Julia Powers, Ventura Lovato, et al. "Food Insecurity and Household Eating Patterns Among Vulnerable American-Indian Families: Associations with Caregiver and Food Consumption Characteristics." *Public Health Nutrition* 16, no.4 (2013): 752–60.

Narveson, Jan. "Property Rights: Original Acquisition and Lockean Provisos." *Public Affairs Quarterly* 13 (1999): 205–27.

Ness, A. R. and J. W. Powles. "Fruit and Vegetables and Cardiovascular Disease: A Review." *International Journal of Epidemiology* 26 (1997): 1–13.

Nestle, Marion. *Food Politics: How the Food Industry Influences Nutrition and Health*. Los Angeles: University of California Press, 2007.

Neuman, William. "F.D.A. and Dairy Industry Spar Over Testing of Milk." Accessed July 28, 2013. www.nytimes.com/2011/01/26/business/26/milk/html.

Nevins, Joseph. *Operation Gatekeeper and Beyond: The War on "Illegals" and the Remaking of the U.S.–Mexico Boundary*. New York: Routledge, 2010.

Newcombe, Mark A., Mary B. McCarthy, James M. Cronin, and Sinead N. McCarthy. "'Eat Like a Man': A Social Constructionist Analysis of the Role of Food in Men's Lives." *Appetite* 59 (2012): 391–98.

Ng, Francis, and M. Ataman Aksoy. *Who are the Net Food Importing Countries?* Washington, D.C.: World Bank, 2008. Accessed March 11, 2015. http://elibrary.worldbank.org/doi/pdf/10.1596/1813-9450-4457.

Nicastro, Juan. "FAO Accepts to Debate Food Sovereignty." *Eurasia Review* (2012): ISSN 2330-717X.

No Reservations. Directed by Scott Hicks. 2007. Burbank, CA: Warner Home Video. 2008. DVD.

Norfolk, Lawrence. "Why Does Nobody Eat in Books?" *The Guardian*. September 6, 2012.

Norton, Bryan G. "Intergenerational Equity and Sustainability." In Bryan G. Norton, *Searching for Sustainability: Interdisciplinary Essays in the Philosophy of Conservation Biology*, 420–55. Cambridge: Cambridge University Press, 2003.

Norton, Bryan G. "What Do We Owe the Future? How Should We Decide?" In Bryan G. Norton, *Searching for Sustainability: Interdisciplinary Essays in the Philosophy of Conservation Biology*, 493–513. Cambridge: Cambridge University Press, 2003. Originally published in *Wolves and Human Communities*, edited by V. A. Sharpe, B. Norton, and S. D. Donnelly. Covelo: Island Press, 2001.

Norton, Bryan G., and Michael A. Toman. "Sustainability: Ecological and Economic Perspectives." In Bryan G. Norton, *Searching for Sustainability: Interdisciplinary Essays in the Philosophy of Conservation Biology*, 225–48. Cambridge: Cambridge University Press, 2003. Originally published in *Land Economics* 73, no. 4 (1997).

Nozick, Robert. *Anarchy, State, and Utopia.* New York: Basic Books, 1974.

Nussbaum, Martha. *Creating Capabilities: The Human Development Approach.* Cambridge, MA: Harvard University Press, 2011.

Nussbaum, Martha C. *Frontiers of Justice: Disability, Nationality, Species Membership.* Cambridge: The Belknap Press of Harvard University Press, 2006.

Nussbaum, Martha. *Women and Human Development: The Capabilities Approach.* Cambridge: Cambridge University Press, 2000.

Nyéléni. "Nyéléni Declaration." *Nyéléni 2007—Forum for Food Sovereignty.* http://nyeleni.org/spip.php?article290.

Nyéléni. "Nyéléni Synthesis Report." *Nyéléni 2007—Forum for Food Sovereignty.* http://nyeleni.org/spip.php?article334.

Ockham, William of. *A Short Discourse on Tyrannical Government.* Translated by John Kilcullen. Cambridge: Cambridge University Press, 1992.

Olson, Christine. "Nutrition and Health Outcomes Associated with Food Insecurity and Hunger." *The Journal of Nutrition* Supplement (Symposium: Advances in Measuring Food Insecurity and Hunger in the U.S.) (1999): 521S–4S.

O'Neill, Catherine. "Variable Justice: Environmental Standards, Contaminated Fish, and 'Acceptable' Risk to Native Peoples." *Stanford Environmental Law Journal* 19 (2000): 3–393.

O'Neill, John, Alan Holland, and Andrew Light. *Environmental Values*. New York: Routledge, 2008.

O'Neill, Onora. *Faces of Hunger: An Essay on Poverty, Justice and Development.* Unwin Hyman, 1986.

Ong Hing, Bill. *Ethical Borders: NAFTA, Globalization, and Mexican Migration.* Philadelphia: Temple University Press, 2010.

Otero, Gerardo, and Cornelia Butler-Flora. "Sweet Protectionism: State Policy and Employment in the Sugar Industries of the NAFTA Countries." In *NAFTA and the Campesinos: The Impact of NAFTA on Small-Scale Agricultural Producers in Mexico and the Prospects for Change*, edited by Juan Rivera, Scott Whiteford, and Manuel Chavez. Scranton: University of Scranton Press, 2008.

Other Worlds. "Uprooting Racism in the Food System: African Americans Organize." Accessed May 1, 2015. http://otherworldsarepossible.org/uprooting-racism-food-system-african-americans-organize.

Panetta, Kasey. "Should You Try a Vegan Diet?" *Men's Health*, June 8, 2012. http://new-mh.menshealth.com/nutrition/vegan-diet-training.

Parasecoli, Fabio. *Food and Men in Cinema: An Exploration of Gender in Blockbuster Movies*. PhD diss., University of Hohenheim, 2009.

Parfit, Derek. *Reasons and Persons*. Oxford: Oxford University Press, 1984.

Patel, Raj. "Food Sovereignty: Power, Gender, and the Right to Food." *PLOS Medicine* 9 (June 26, 2012): 1–4. doi:10.1371/journal.pmed.1001223.

Patel, Raj. "What Does Food Sovereignty Look Like?" *The Journal of Peasant Studies* 36 (2009): 663–706.

Paxton George, Kathryn. "Should Feminists be Vegetarians?" *Signs: Journal of Women in Culture and Society* 19 (1994): 405–34.

PBS. *Black Farming History: The Civil Rights Years (1954–1968)*. Accessed May 4, 2015. http://www.pbs.org/itvs/homecoming/history5.html.

Pearce, J., T. Blakely, K. Witten, and P. Bartie. "Neighborhood Deprivation and Access to Fast-Food Retailing: A National Study." *American Journal of Preventative Medicine* 32 (2007): 375–82.

Pechlaner, Gabriela, and Gerardo Otero. "Neoliberalism and Food Vulnerability: The Stakes for the South." In *Food Security, Nutrition and Sustainability*, edited by Geoffrey Lawrence, Kristen Lyons, and Tabatha Wallington. Sterling, VA: Earthscan, 2010.

Peplau, Letitia Anne, David A. Frederick, Curtis Yee, Natalya Maisel, Janet Lever, and Negin Ghavami. "Body Image Satisfaction in Heterosexual, Gay and Lesbian Adults." *Archives of Sexual Behavior* 38 (2009): 713–25.

Physicians Committee on Responsible Medicine. "Meat Consumption and Cancer Risk." Accessed May 14, 2014. http://www.pcrm.org/health/cancer-resources/diet-cancer/facts/meat-consumption-and-cancer-risk.

Physicians Committee on Responsible Medicine. "Meat Week Should Be Renamed Erectile Dysfunction Acceptance Week, say Doctors." Accessed July 14, 2014. http://www.pcrm.org/media/news/meat-week-be-renamed-erectile-dysfunction-week.

Pike, Kathleen M., and Amy Borovoy. "The Rise of Eating Disorders in Japan: Issues of Culture and Limitations of the Model of 'Westernization.'" *Culture, Medicine, and Psychiatry* 28 (2004): 493–531.

Pilgrim, Karyn. "'Happy Cows,' 'Happy Beef': A Critique of the Rationales for Ethical Meat." *Environmental Humanities* 3 (2013): 111–27.

Pimbert, Michel. *Towards Food Sovereignty: Reclaiming Autonomous Food Systems*. London: International Institute for Environment and Development, 2009.

Pimbert, Michel. "Women and Food Sovereignty." *LEISA Magazine* 25 (2009): 6–9.

"Plastic Makes Perfect." *Economist Online*, January 30, 2013. Accessed April 22, 2015. http://www.economist.com/blogs/graphicdetail/2013/01/daily-chart-22.

Pogge, Thomas. "'Assisting' the Global Poor." In *The Ethics of Assistance: Morality and the Distant Needy*, edited by Deen K. Chatterjee, 260–88. Cambridge: Cambridge University Press, 2004.

Pogge, Thomas. *World Poverty and Human Rights*. Cambridge: Polity Press, 2008.

Pollan, Michael. *The Omnivore's Dilemma: A Natural History of Four Meals*. New York: Penguin Books, 2006.

Powell, Lisa M., Sandy Slater, Donka Mirtcheva, Yanjun Bao, and Frank J. Chaloupka. "Food Store Availability and Neighborhood Characteristics in the United States." *Preventive Medicine* 44 (2007): 189–95.

Project Concern International. "Women's Empowerment and Poverty." http://www.pciglobal.org/womens-empowerment-poverty/.

Rabelais, Francois. *Pantagruel, Book 4–5*. Translated by William Francis Smith. London: Watt, 1893.

Race Forward: The Center for Racial Justice Innovation. "Food Justice." Accessed May 4, 2015. https://www.raceforward.org/research/reports/food-justice.

Ranco, Darren J., Catherine A. O'Neill, Jamie Donatuto, and Barbara L. Harper. "Environmental Justice, American Indians and the Cultural Dilemma: Developing Environmental Management for Tribal Health and Well-Being." *Environmental Justice* 4 (2011): 221–30.

Ransom, Elizabeth, and Leslie K. Elder. "Nutrition of Women and Adolescent Girls: Why It Matters." Population Reference Bureau. Accessed March 8, 2015. http://www.prb.org/Publications/Articles/2003/NutritionofWomenandAdolescentGirls-WhyItMatters.aspx.

Rapport, David J. "Ecosystem Health: More than a Metaphor?" *Environmental Values* 4 (1995): 287–309.

Rapport, David J. "Sustainability Science: An Ecohealth Perspective." *Sustainability Science* 2 (2007): 77–84.

Rawls, John. *Justice as Fairness: A Restatement*, edited by Erin Kelly. Cambridge, MA: Harvard University Press, 2001.

Rawls, John. *Political Liberalism*. New York: Columbia University Press, 1993.

Rawls, John. *A Theory of Justice*. Revised ed. Cambridge, MA: The Belknap Press of Harvard University Press, 1999.

Read, Piers Paul. *Alive: The Story of the Andes Survivors*. J. B. Lippincott, 1974.

The Recipe. Directed by Anna Lee. 2010. Seoul: Film It Suda.

Regan, Tom. *The Case for Animal Rights*. Berkeley: University of California Press, 2004.

Reisig, Vmt, and A. Hobbiss. "Food Deserts and How to Tackle Them: A Study of One City's Approach." *Health Education Journal* 59 (2000): 137–48.

Rickless, Samuel C. *Locke*. Malden, MA: John Wiley & Sons, 2014.

Riffkin, Rebecca. "Mississippians' Struggles to Afford Food Continued in 2013." *Gallup*, March 21, 2014. Accessed March 11, 2015. http://www.gallup.com/poll/167774/mississippians-struggles-afford-food-continue-2013.aspx.

Riverwest Co-op. "Homepage." Accessed April 9, 2013. www.riverwestcoop.org.

Rivera, Juan. "Multinational Agribusiness and Small Corn Producers in Rural Mexico: New Alternatives for Agricultural Development." In *NAFTA and the Campesinos: The Impact of NAFTA on Small-Scale Agricultural Producers in Mexico and the Prospects for Change*, edited by Juan Rivera, Scott Whiteford, and Manuel Chavez. Scranton: University of Scranton Press, 2008.

Rivera, Juan, and Scott Whiteford. "Mexican Agriculture and NAFTA—Prospects for Change." In *NAFTA and the Campesinos: The Impact of NAFTA on Small-Scale Agricultural Producers in Mexico and the Prospects for Change*, edited by Juan

Rivera, Scott Whiteford, and Manuel Chavez. Scranton: University of Scranton Press, 2008.

Robin, Marie-Monique. *The World According to Monsanto: Pollution, Corruption, and the Control of Our Food Supply*. Translated by George Holoch. New York: New Press, 2009.

Rocheleau, Diane, Barbara Thomas-Slayter, and Esther Wangari, editors. *Feminist Political Ecology: Global Issues and Local Experience*. New York: Routledge, 1996.

Roos, Gun, and Margareta Wandel. "'Because I'm Hungry, Because It's Good, and to Become Full': Everyday Eating Voiced by Carpenters, Drivers, and Engineers in Contemporary Oslo." *Food and Foodways* 13 (2005): 169–280.

Rop, Rosemary. "Mainstreaming Gender in Water and Sanitation: Gender in Water and Sanitation." World Bank Water and Sanitation Program working paper, 2010. Accessed March 8, 2015. http://www.wsp.org/sites/wsp.org/files/publications/WSP-gender-water-sanitation.pdf.

Rose-Jacobs, Ruth, Maureen M. Black, Patrick H. Casey, John T. Cook, Diana B. Cutts, Mariana Chilton, Timothy Heeren, Suzette M. Levenson, Alan F. Meyers, and Deborah A. Frank. "Household Food Insecurity: Associations With At-Risk Infant and Toddler Development." *Pediatrics* 121 (2008): 65–72.

Rothgerber, Hank. "Real Men Don't Eat (Vegetable) Quiche: Masculinity and the Justification of Meat Consumption." *Psychology of Men and Masculinity* 14 (2012): 363–75.

Ruby, Matthew B., and Steven J. Heine. "Meals, Morals, and Masculinity." *Appetite* 56 (2011): 447–50.

Rudy, Kathy. "Locavores, Feminism, and the Question of Meat." *The Journal of American Culture* 35 (2012): 26–36.

Runge, C. Ford, and Benjamin Senauer. "A Removable Feast." In *The Ethics of Food: A Reader for the 21st Century*, edited by Gregory E. Pence, 180–90. New York: Rowman and Littlefield, 2002.

Ruttan, Vernon W. *Why Food Aid?* Baltimore, MD: Johns Hopkins University Press, 1993.

Sachs, Carolyn. "Feminist Food Sovereignty: Crafting a New Vision." Yale University, 2013. http://www.yale.edu/agrarianstudies/foodsovereignty/pprs/58_Sachs_2013.pdf.

Sasson, Albert. "Food Security for Africa: An Urgent Global Challenge." *Agriculture and Food Security* 1 (2012) doi:10.1186/2048-7010-1-2.

Sbicca, Joshua. "Growing Food Justice by Planting an Anti-Oppressive Foundation: Opportunities and Obstacles for a Budding Social Movement." *Agriculture and Human Values* 29 (2012): 455–66.

Scanlon, Thomas. "Nozick on Rights, Liberty, and Property." *Philosophy and Public Affairs* 6 (1976): 3–25.

Schanbacher, William D. *The Politics of Food: The Global Conflict Between Food Security and Food Sovereignty*. Santa Barbara, CA: Praeger, 2010.

Schlosberg, David. "Reconceiving Environmental Justice: Global Movements and Political Theories." *Environmental Politics* 13 (2004): 517–40.

Schlosberg, David, and David Carruthers. "Indigenous Struggles, Environmental Justice, and Community Capabilities." *Global Environmental Politics* 10 (2010): 12–35.

Schlosser, Eric. *Fast Food Nation: The Dark Side of the All-American Meal*. New York: Houghton Mifflin, 2002.

Schuett, Kat. "Guayaki Invites You to 'Share the Gourd' to Empower Indigenous Communities." *Fair World Project*. Fairworldproject.org (2014).

Schultz, E. J. "Weight Watchers Pick a New Target: Men." *Advertising Age*. Last modified April 22, 2011. http://adage.com/article/news/weight-watchers-picks-a-target-men/227155/.

Schultz, Theodore W. "Value of US Farm Surpluses to Underdeveloped Countries." *Journal of Farm Economics* 42 (1960): 1019–30.

Scicluna, Henry. "The Health Situation of the Roma in Europe." 4th Conference on Migrant and Ethnic Minority Health in Europe, June 21–23, 2013, Universita Bocconi, Milan, Italy.

Scoville, J. Michael. "Historical Environmental Values." *Environmental Ethics* 35 (2013): 7–25.

Searchinger, Tim, Craig Hanson, Janet Ranganathan, Brian Lipinski, Richard Waite, Robert Winterbottom, Ayesha Dinshaw, and Ralph Hemlich. "Creating a Sustainable Future: Interim Findings." World Resources Institute, 2013. Accessed July 7, 2014. http://www.wri.org/publication/creating-sustainable-food-future-interim-findings.

Second Livestock. "Virtual Free Range." http://www.secondlivestock.com/public/vfr.php.

Seeking a Friend for the End of the World. Directed by Lorene Scafaria. 2012. Universal City, CA: Focus Features. 2012. DVD.

Sen, Amartya. *Development as Freedom*. New York: Oxford University Press, 1999.

Sen, Amartya. *Poverty and Famines: An Essay on Entitlement and Deprivation*. New York: Oxford University Press, 1983.

Sen, Amartya. *Poverty and Famines*. Oxford: Oxford University Press, 1981.

Sex and the City: The Movie. Directed by Michael Patrick King 2008. Burbank, CA: New Line Home Video. DVD.

Sex and the City 2. Directed by Michael Patrick King. 2010. Burbank, CA: New Line Home Video. DVD.

Shallow Hal. Directed by Bobby Farrelly and Peter Farrelly. 2001. Los Angeles: 20th Century Fox. 2002. DVD.

Shaw, Hillary J. *The Consuming Geographies of Food*. New York: Routledge, 2014.

Shepherd, Benjamin. "Thinking Critically about Food Security." *Security Dialogue* 43 (2012): 195–212.

Shiva, Vandana. "Globalization and the War against Farmers and the Land." In *The Essential Agrarian Reader: The Future of Culture, Community, and the Land*, edited by Norman Wirzba, 121–39. Lexington: University Press of Kentucky, 2003.

Shiva, Vandana. "The Impoverishment of the Environment: Women and Children Last." In *Environmental Philosophy: From Animal Rights to Radical Ecology* (4th ed.), edited by Michael E. Zimmerman, J. Baird Callicott, Karen J. Warren, Irene

J. Klaver, and John Clark, 178–93. Upper Saddle River: Pearson Education, Inc., 2005. Originally published in Maria Mies and Vandana Shiva, *Ecofeminism*. Atlantic Highlands, NJ: Zed Books, 1993.

Shiva, Vandana. *Staying Alive: Women, Ecology and Development*. Zed Books, 1988.

Shiva, Vandana. *Stolen Harvest: The Hijacking of the Global Food Supply*. Cambridge: South End Press, 2000.

Shrader-Frechette, K. S. *Environmental Justice: Creating Equality, Reclaiming Democracy*. New York: Oxford University Press, 2002.

Shrader-Frechette, K. S. *Risk and Rationality: Philosophical Foundations for Populist Reforms*. Berkeley, CA: University of California Press, 1991.

Simmons, A. John. "Historical Rights and Fair Shares." *Law and Philosophy* 14 (1995): 149–84.

Simmons, A. John. *The Lockean Theory of Rights*. Princeton: Princeton University Press, 1992.

Simply Irresistible. Directed by Mark Tarlov. 1999. Los Angeles: 20th Century Fox. 2002. DVD.

Singer, Peter. *Animal Liberation*. New York: Harper Collins, 2009.

Singer, Peter. "Famine, Affluence, and Morality." *Philosophy and Public Affairs* 1 (1972): 229–43.

Singer, Peter. *The Life You Can Save*. New York: Random House, 2009.

Singer, Peter. *Practical Ethics*. 3rd ed. Cambridge: Cambridge University Press, 2011.

Singer, Peter, and Jim Mason. *The Way We Eat: Why Our Food Choices Matter*. USA: Rodale, Inc., 2006.

The Single Mom's Club. Directed by Tyler Perry. 2014. Santa Monica, CA: Lion's Gate. DVD.

Slocum, Rachel. "Race in the Study of Food." *Progress in Human Geography* 35 (2011): 303–27.

Snowdon D. A., R. L. Phillips, and G. E. Fraser. "Meat Consumption and Fatal Ischemic Heart Disease." *Preventive Medicine* September, 13 (1984): 490–500.

Sobel, Jeffery. "Men, Meat, and Marriage: Models of Masculinity." *Food and Foodways* 13 (2005): 135–58.

Solovay, Sondra. *Tipping the Scales of Justice: Fighting Weight-Based Discrimination*. New York: Prometheus Books, 2000.

Solow, Robert. "Sustainability: An Economist's Perspective." In *Economics of the Environment: Selected Readings* (3rd ed.), edited by Robert Dorfman and Nancy Dorfman, 179-87. New York: W. W. Norton and Company, 1993.

Something New. Directed by Sanaa Hamri. 2006. Universal City, CA: Focus Features. 2006. DVD.

Sreenivasan, Gopal. *The Limits of Lockean Rights in Property*. New York: Oxford University Press, 1995.

Stanescu,Vasile. "'Green' Eggs and Ham? The Myth of Sustainable Meat and the Danger of the Local." *Journal for Critical Animal Studies* 8 (2010): 8–32.

Sterba, James P. "From Biocentric Individualism to Biocentric Pluralism." *Environmental Ethics* 17 (1995): 191–207.

Sterba, James P. "Global Justice for Humans or For All Living Beings and What Difference It Makes." *The Journal of Ethics* 9 (2005): 283–300.

Sterba, James P. *How to Make People Just: A Practical Reconciliation of Alternative Conceptions of Justice*. Totowa, NJ: Rowman & Littlefield Publishers, 1988.

Sterba, James P. "The Welfare Rights of Distant Peoples and Future Generations: Moral Side Constraints on Social Policy." *Social Theory and Practice* 7 (1981): 99–119.

Stern, Paul. C., and Harvey V. Fineberg, editors. *Understanding Risk: Informing Decisions in a Democratic Society.* Washington, D.C.: National Academy Press, 1996.

Stevenson, Jim. "Dietary Influences on Cognitive Development and Behavior in Children." *Proceedings of the Nutrition Society* 65 (2006): 361–65.

Stibble, Arran. "Health and the Social Construction of Masculinity in *Men's Health* Magazine." *Men and Masculinities* 7 (2004): 31–51.

Stiglitz, Joseph E. *Making Globalisation Work.* New York: Norton, 2006.

Suagee, Dean B. "Turtle's War Party: An Indian Allegory on Environmental Justice." *Journal of Environmental Law and Litigation* 9 (1994): 461–97.

Swami, Viren, et al., "The Attractive Female Body Weight and Female Body Dissatisfaction in 26 Countries across 10 World Regions: Results of the International Body Project I." *Personality and Social Psychology Bulletin* 36 (2010): 309–25.

Sweeting, H., and P. West. "Gender Differences in Weight Related Concerns in Early to Late Adolescence." *Journal of Epidemiology and Community Health* 56 (2002): 700–701.

Sweet Water Foundation. "Home." Accessed April 9, 2013. www.sweetwater-organic.com.

Swinburne, Richard. *Providence and the Problem of Evil.* Oxford: Oxford University Press, 1998.

Takaki, Ronald. *A Different Mirror: A History of Multicultural America.* New York: Back Bay Books, 2008.

Tammy. Directed by Ben Falcone. 2014. Burbank, CA: New Line Cinema. DVD.

Tarnapol Whitacre, Paula, Peggy Tsai, and Janet Mulligan. "The Public Health Effects of Food Deserts: Workshop Summary." Washington, D.C.: National Academy of Sciences, 2009.

Taylor, Chole. "Foucault and the Ethics of Eating." *Foucault Studies* 9 September (2010): 71–88.

Taylor, Dorceta E. "The Rise of the Environmental Justice Paradigm: Injustice Framing and the Social Construction of Environmental Discourses." *American Behavioral Scientist* 43 (2000): 508–80.

Tharyan P., and G. Gopalakrishanan. "Erectile Dysfunction." *Clinical Evidence* 1803 (2006). Accessed June 12, 2014. http://www.ncbi.nlm.nih.gov/pmc/articles/PMC2907627/.

This is 40. Directed by Judd Apatow. 2012. Universal City: Universal Pictures. 2013. DVD.

Thompson, Allen. "Radical Hope for Living Well in a Warmer World." *Journal of Agricultural and Environmental Ethics* (2009) DOI 10.1007/s10806-009-9185-2.

Thompson, Cheryl. "Neoliberalism, Soul Food, and the Weight of Black Women." *Feminist Media Studies* (2015): 1–21. DOI: 10.1080/14680777.2014.1003390.

Thompson, Paul B. "Food Aid and the Famine Relief Argument (Brief Return)." *The Journal of Agricultural and Environmental Ethics* 23 (2010): 209–27.

Thurow, Roger, and Scott Kilman. *Enough: Why the World's Poorest Starve in an Age of Plenty*. New York: Public Affairs, 2009.

Tobin, K. J. "Fast-Food Consumption and Educational Test Scores in the USA." *Child: Care, Health, and Development* 39 (2013): 118–24.

Today's Special. Directed by David Kaplan. 2009. Niwot, CO: Flatiron Films. 2012. DVD.

Tortilla Soup. Directed by Maria Ripoll. 2001. Culver City: Sony Pictures Home Entertainment. 2002. DVD.

Tully, Jarries. *A Discourse on Property: John Locke and His Adversaries*. Cambridge: Cambridge University Press, 1980.

UNICEF. "Global Malnutrition Trends (1990–2013)." Accessed March 11, 2015. http://data.unicef.org/resources/2013/webapps/nutrition.

UNICEF. "Undernutrition Contributes to Half of all Deaths in Children Under 5 and is Widespread in Asia and Africa." Last updated February 2015. http://data.unicef.org/nutrition/malnutrition.

United Nations. "Goal 3: Promote Gender Equality and Empower Women." http://www.un.org/millenniumgoals/gender.shtml.

United Nations. "The Millenium Development Goals Report: Gender Chart." (2012) Accessed March 8, 2015. http://www.unwomen.org/~/media/headquarters/attachments/sections/library/publications/2012/12/mdg-gender-web%20pdf.pdf.

United Nations. The Universal Declaration of Human Rights. http://www.un.org/en/documents/udhr/index.shtml#ap.

United Nations Development Programme. "Gender and Poverty Reduction." Accessed March 8, 2015. http://www.undp.org/content/undp/en/home/ourwork/povertyreduction/focus_areas/focus_gender_and_poverty.html.

United Nations Development Programme. "The Millennium Development Goals: Eight Goals for 2015." Accessed April 18, 2015. http://www.undp.org/content/undp/en/home/mdgoverview.html.

UN News Centre. "World Population Projected to Reach 9.6 Billion by 2050—UN Report." 2013. Accessed July 7, 2014. http://www.un.org/apps/news/story.asp?NewsID=45165#.U78ySvk7uSp.

UN Water. "Gender, Water and Sanitation: A Policy Brief." Accessed March 8, 2015. http://www.un.org/waterforlifedecade/pdf/un_water_policy_brief_2_gender.pdf

UN Water. "Water is Food." Accessed March 8, 2015. http://www.unwater.org/worldwaterday/learn/en/?section=c325501.

The Upside of Anger. Directed by Mike Bender. 2005. Los Angeles: New Line Home Cinema. 2010. DVD.

Urrea, Luis Alberto. *The Devil's Highway*. New York: Back Bay Books, 2005.

"The US—Mexico Border: Secure Enough," *The Economist*, June 22, 2013. http://www.economist.com/news/united-states/21579828-spending-billions-more-fences-and-drones-will-do-more-harm-good-secure-enough.

United States Bureau of Labor Statistics, 2014. http://www.bls.gov/cps/cpsaat11.htm.

United States Department of Agriculture. "Food Access Research Atlas." Accessed November 20, 2014. www.ers.usda.gov/data-products/food-access-research-atlas/go-to-the-atlas.aspx.

United States Department of Agriculture. "Food Deserts." Accessed 3 June, 2013. http://apps.ams.usda.gov/fooddeserts/foodDeserts.aspx

United States Department of Health and Human Services. "Community Economic Development Healthy Food Financing Initiative Projects." Accessed February 24, 2015. https://www.acf.hhs.gov/hhsgrantsforecast/index.cfm?switch=grant.view&gff_grants_forecastInfoID=67242.

United States Department of Labor. "Occupational Employment Statistics." Accessed May 4, 2015. http://www.bls.gov/oes/current/oes452099.htm.

Vivas, Esther. "Without Women There Is No Food Sovereignty." *International Viewpoint*, February 2, 2012.

Waitress. Directed by Adrienne Shelly. 2007. Los Angeles: Fox Searchlight. DVD.

Waldron, Jeremy. *Law and Disagreement*. Oxford: Oxford University Press, 1999.

Waldron, Jeremy. *The Right to Private Property*. Oxford: Clarendon Press, 1988.

Walker, Renee E., Christopher R. Keane, and Jessica G. Burke. "Disparities and Access to Healthy Food in the United States: A Review of Food Deserts Literature." *Health & Place* 16, no.8 (2010): 876–84.

Water.org. "Millions Lack Safe Water." http://water.org/water-crisis/water-facts/water.

Weatherspoon, Dave, J. Oehmke, A. Dembélé, M. Coleman, T. Satimanon, and L. Weatherspoon. "Price and Expenditure Elasticities for Fresh Fruits in an Urban Food Desert." *Urban Studies* 50 (2013): 88–106.

Weinreb, Linda, Cheryl Wehler, Jennifer Perloff, Richard Scott, David Hosmer, Linda Sagor, and Craig Gundersen. "Hunger: Its Impact on Children's Health and Mental Health." *Pediatrics* 110 (2002): e41.

Wekerle, Gerda R. "Food Justice Movements: Policy, Planning, and Networks." *Journal of Planning Education and Research* 23 (2004): 378–86.

Wells, Michael. "Biodiversity Conservation, Affluence and Poverty: Mismatched Costs and Benefits and Efforts to Remedy Them." *Ambio* 21 (1992): 237–43.

Werkheiser, Ian and Samantha Noll. "From Food Justice to a Tool of the Status Quo: Three Sub-Movements within Local Food." *Journal of Agricultural and Environmental Ethics* 27 (2014): 201–10.

Wheelan, Charles. *Naked Economics: Undressing the Dismal Science*. New York: W.W. Norton & Company, 2010.

White, Marceline, Sarah Gammage, and Carlos Salas Paez. "NAFTA and the FTAA: Impact on Mexico's Agricultural Sector." IATP Institute for Agriculture and Trade Policy, 2003. Accessed July 10, 2014. http://www.iatp.org/files/NAFTA_and_the_FTAA_Impact_on_Mexicos_Agricultu.pdf.

Whyte, Kyle P. "Food Justice and Collective Food Relations." Forthcoming in *The Ethics of Food: An Introductory Textbook*, edited by Anne Barnhill, Mark Budolfson, and Tyler Doggett. Oxford: Oxford University Press, 2015. Page references are to the draft available at: http://www.academia.edu/10326764/Food_Justice_and_Collective_Food_Relations_for_Introductory_Food_Ethics_Textbook.

Widerquist, Karl. "Lockean Theories of Property: Justification for Unilateral Appropriation." *Public Reason* 2 (2010): 3–26.

Wiggins, David. *Needs, Values, Truth: Essays in the Philosophy of Value*. 3rd ed. Oxford: Oxford University Press, 1998.

Williams, Bernard. "Must a Concern for the Environment be Centred on Human Beings?" In Bernard Williams, *Making Sense of Humanity and Other Philosophical Papers*, 233–40. Cambridge: Cambridge University Press, 1995. Originally published in *Ethics and the Environment*. Edited by C. C. W. Taylor. Oxford: Corpus Christi College, 1992.

Wittman, Hannah, Annette Desmarais, and Nettie Wiebe, editors. *Food Sovereignty: Reconnecting Food, Nature, and Community*. Oakland, CA: Food First Books, 2010.

Wittman, Hannah, Annette Desmarais, and Nettie Wiebe. "The Origins and Potential of Food Sovereignty." In *Food Sovereignty: Reconnecting Food, Nature, and Community*, edited by Wittman, Desmarais, and Wiebe, 1–14. Oakland, CA: Food First, 2010.

Woman on Top. Directed by Fina Torres. 2000. Los Angeles: Fox Searchlight. 2003. DVD.

World Bank. "Nutrition at a Glance: Guatemala." Accessed March 9, 2015. http://siteresources.worldbank.org/NUTRITION/Resources/281846-1271963823772/Guatemala.pdf.

World Food Programme. *WFP Gender Policy*. Rome: WFP, 2009. Accessed March 10, 2015. http://documents.wfp.org/stellent/groups/public/documents/communications/wfp203758.pdf.

World Food Programme. "Who Are the Hungry?" Accessed March 25, 2015. http://www.wfp.org/hunger/who-are.

World Health Organization. "Food Security." Accessed July 15, 2014. http://www.who.int/trade/glossary/story028/en/.

Wrigley, Neil. "Food Deserts in British Cities: Policy Context and Research Priorities." *Urban Studies* 39 (2002): 2029–40.

Xu, Xiaoyan, David Mellor, Melanie Kiehne, Lina A. Ricciardelli, Marita P. McCabe, and Yangang Xu. "Body Dissatisfaction, Engagement in Body Change Behaviors and Sociocultural Influences on Body Image among Chinese Adolescents." *Body Image* 7 (2010): 156–64.

Yang, Jie. "*Nennu* and *Shunu*: Gender, Body Politics and the Beauty Economy of China." *Signs* 36 (2011): 333–57.

Young, Iris Marion. *Justice and the Politics of Difference*. Princeton: Princeton University Press, 1990.

Zedillo Ponce de Leon, Ernesto. "NAFTA at 20." *Americas Quarterly*, 2014.

Ziegler, Jean. *Betting on Famine: Why the World Still Goes Hungry*. New York: New Press, 2013.

Zimmerman, Steve. *Food in the Movies*, 2nd edition. Jefferson, NC: McFarland & Company, Inc., 2010. Kindle Edition.

Index

About the Contributors

Margaret Crouch is a professor of philosophy at Eastern Michigan University. She published *Thinking about Sexual Harassment: A Guide for the Perplexed* (Oxford University Press, 2001). Her recent publications include "Implicit Bias and Gender (and other sorts of) Diversity in Philosophy and the Academy in the Context of the Corporatized University" in a special issue of the *Journal of Social Philosophy* which she coedited (vol. 43, 2012) and "Sexual Harassment in Public Places" (*Social Philosophy Today: Gender, Equality and Social Justice,* vol. 25, Philosophy Documentation Center, 2009). She is editor of the *APA Newsletter on Philosophy and Feminism*, treasurer of the Association of Feminist Ethics and Social Theory, and President of the North American Society for Social Philosophy. She teaches interdisciplinary courses in environmental studies, film, philosophy, and gender.

J. M. Dieterle is a professor of philosophy at Eastern Michigan University. Her current research involves food justice and food ethics. Her most recent publications include "Affording Disaster: Concealed Carry on Campus" (coauthored with W. John Koolage) in *Public Affairs Quarterly* and "Unnecessary Suffering" in *Environmental Ethics.* She has additional published articles in *Public Affairs Quarterly*, *Bioethics*, *Journal of Business Ethics, Philosophia Mathematica,* and *Erkenntnis.* Her teaching interests include metaphysics, epistemology, logic, and food ethics.

Liz Goodnick is an assistant professor of philosophy at Metropolitan State University of Denver. She received her PhD from the University of Michigan in 2010. While her research has primarily focused on early modern philosophy and the philosophy of religion, her personal commitment to animal rights has recently given rise to a professional interest in the philosophy of food.

In her spare time, she enjoys vegetarian cooking and hiking in the beautiful Rocky Mountains with her husband and their dog Banjo.

Stephen Minister is the Clara Lea Olsen Professor of Ethical Values and an assistant professor of philosophy at Augustana College in South Dakota. He specializes in ethics and continental philosophy, with particular interest in issues related to global poverty, human rights, and social justice. He is the author of *De-Facing the Other: Reason, Ethics, and Politics after Difference* (Marquette University Press, 2012) as well as more than a dozen journal articles and book chapters. He regularly teaches study-abroad courses in Mexico, Guatemala, and Cuba.

Mark Navin is Associate Professor of Philosophy at Oakland University (Rochester, MI). His research is primarily in social and political philosophy. His book *Values and Vaccine Refusal* was published by Routledge in 2015, and his articles have appeared in journals including *Social Theory and Practice*, *Public Affairs Quarterly*, *Kennedy Institute of Ethics Journal*, *Ethical Theory and Moral Practice*, and *Journal of Agricultural and Environmental Ethics*.

J. Michael Scoville is an assistant professor of philosophy at Eastern Michigan University. His primary research interests are environmental ethics and ethical theory. In 2012, he received the Holmes Rolston III Early Career Essay Prize for his article "Historical Environmental Values" (published in *Environmental Ethics*, Spring 2013). His current work is focused on two main projects. One is assessing different conceptions of sustainability. The other is exploring possible defenses of biological or ecological integrity (or "integrity" for short) as a conservation concept. Integrity refers to a property of landscapes that are relatively unmodified by human activity and that have their native biota largely intact. The defense of integrity faces a number of challenges given pervasive human modifications of nature, disagreement about the normative significance of native species, and questions concerning the relative importance of preserving or restoring integrity.

Nancy E. Snow is Professor of Philosophy and Director of the Institute for the Study of Human Flourishing at the University of Oklahoma. Her research interests are in virtue ethics and moral psychology. Her book, *Virtue as Social Intelligence: An Empirically Grounded Theory* (Routledge, 2010), argues for the empirical adequacy of virtues as traditionally conceived in philosophy. She continues to write on empirical aspects of the psychology of virtue, has written a book-length manuscript on hope, and is at work on a book on virtue ethics and virtue epistemology. A coedited volume with Franco V. Trivigno, *The Philosophy and Psychology of Character and Happiness*, was published by Routledge (2014), and an edited volume, *Cultivating Virtue: Perspectives* *from Philosophy, Theology, and Psychology*, by Oxford University Press (2014).

Jennifer Szende is a postdoctoral fellow in the environmental axis at the Centre de Recherche en Éthique at l'Université de Montréal. She completed her PhD at Queen's University, Canada, in 2013. She works on issues of justice in international relations, environmental justice, and human rights. Her research project at the Centre de Recherche en Éthique aims to build a bridge between global justice and environmental justice. Her research has appeared in the *Journal of Global Ethics* and the *Encyclopedia of Global Justice* edited by Deen Chatterjee.

Steve Tammelleo is an assistant professor of philosophy at the University of San Diego. He completed his PhD at the University of Memphis. Steve has published work on Hispanic/Latino identity, philosophy of mind, and philosophy of language. He enjoys teaching classes in logic, business ethics, Latin American thought, and introduction to philosophy. Steve's current research focuses on Foucault, border policy, immigration, and ethnic identity. Steve has traveled extensively in Latin America. Steve enjoys yoga, biking, hiking, and volleyball.

Paul B. Thompson is the W. K. Kellogg Chair in Agricultural, Food, and Community Ethics at Michigan State University. He is the author of thirteen books and editions, such as *The Spirit of the Soil: Agriculture and Environmental Ethics; The Ethics of Aid and Trade; Food Biotechnology in Ethical Perspective*, and is coeditor of *The Agrarian Roots of Pragmatism.* He has served on many national and international committees on agricultural biotechnology and contributed to the National Research Council report "The Environmental Effects of Transgenic Plants." He is a past president of the Agriculture, Food, and Human Values Society and the Society for Philosophy and Technology, and is secretary of the International Society for Environmental Ethics. He has continuing interests in environmental and agricultural ethics.

Shakara Tyler is a PhD student at Michigan State University in the Department of Community Sustainability. She currently engages in participatory action research with underserved farming communities and utilizes critical (race) theories and decolonial theories to assess and facilitate capacity building and resistance toward the community-identified goals of food sovereignty, particularly food sovereignty for Afro-descendent communities throughout the African Diaspora.

Lori Watson is an associate professor of philosophy and Chair of the Philosophy Department at the University of San Diego. Professor Watson's areas of research include political philosophy, feminist theory, and philosophy of law.

Her primary area of research is concerned with articulating a theory of justice that is responsive to conditions of reasonable pluralism that constitute modern democracies while at the same time securing substantive gender equality. She has published many articles on this topic and is working on a book-length project with Dr. Christie Hartley (GSU).

Ian Werkheiser is Assistant Professor of Philosophy at the University of Texas Rio Grande Valley. His work focuses on environmental justice, food sovereignty, and the social and political dimensions of epistemology. In particular, he is interested in the ways communities respond to problems in the environment and food systems in the context of oppression or marginalization. Ian is currently working with La Via Campesina on a project looking at barriers to women's participation in the food sovereignty movement.

Nancy Williams is an associate professor of philosophy at Wofford College in Spartanburg, South Carolina. Her main research interests are feminist philosophy and ethical theory with a particular focus on animal protectionism and food ethics. Her most recent publication, "The Ethics of Care and Humane Meat: Why Care Is Not Ambiguous about 'Humane' Meat," will appear in the *Journal of Social Philosophy* (forthcoming spring 2015). Other publications include "Affected Ignorance and Animal Suffering: Why the Lack of Extensive Debate about Factory Farming May Put Us at Moral Risk," in the *Journal of Agricultural and Environmental Ethics* (August 2008) and "Feminist Ethics," with Rosemarie Tong, *The Stanford Encyclopedia of Philosophy* (Last revised May 2009), Edward N. Zalta (ed.) (http://plato.stanford.edu/archives/win2003/entries/feminism-ethics/). During the school year she enjoys teaching courses on the philosophy of food, ethical veganism, and environmental ethics but her advocacy sends her west in the summer to volunteer at various animal sanctuaries. She is also chair of the Board of Directors for a nonprofit low-cost spay and neuter clinic in Spartanburg.

Lightning Source UK Ltd.
Milton Keynes UK
UKHW011920011118
331594UK00001B/91/P

9 781783 483877